THE COMMONWEALTH AND
Joint Chairmen of the Honorary
SIR ROBERT ROBINSON, O.
DEAN ATHELSTAN SPILF
Publisher: ROBERT MAX

MATHEMATICS DIVISION
General Editors: W. J. LANGFORD, E. A. MAXWELL

MODERN MATHEMATICS
IN SECONDARY SCHOOLS

MODERN MATHEMATICS IN SECONDARY SCHOOLS

D. T. E. MARJORAM

THE QUEEN'S AWARD
TO INDUSTRY 1966

PERGAMON PRESS

OXFORD · LONDON · EDINBURGH · NEW YORK
TORONTO · SYDNEY · PARIS · BRAUNSCHWEIG

Pergamon Press Ltd., Headington Hill Hall, Oxford
4 & 5 Fitzroy Square, London W.1

Pergamon Press (Scotland) Ltd., 2 & 3 Teviot Place, Edinburgh 1

Pergamon Press Inc., Maxwell House, Fairview Park, Elmsford,
New York 10523

Pergamon of Canada, Ltd., 207 Queen's Quay West, Toronto 1

Pergamon Press (Aust.) Pty. Ltd., 19a Boundary Street,
Rushcutters Bay, N.S.W. 2011, Australia

Pergamon Press S.A.R.L., 24 rue des Écoles, Paris 5ᵉ

Vieweg & Sohn GmbH, Burgplatz 1, Braunschweig

Copyright © 1964 Pergamon Press Ltd.

All Rights Reserved. No part of this publication may be reproduced, stored in a retrieval system, or transmitted, in any form or by any means, electronic, mechanical, photocopying, recording or otherwise, without the prior permission of Pergamon Press Ltd.

First edition 1964   Reprinted 1965

Reprinted (with corrections) 1966

Reprinted 1969

Library of Congress Catalog Card No. 64-21690

Printed in Great Britain by A. Wheaton & Co., Exeter

This book is sold subject to the condition
that it shall not, by way of trade, be lent,
resold, hired out, or otherwise disposed
of without the publisher's consent,
in any form of binding or cover
other than that in which
it is published.

08 010718 4 (flexicover)
08 010719 2 (hard cover

ACKNOWLEDGEMENTS

THIS book has been written at a time when the content, methods and aims of mathematical education at all levels are undergoing a searching re-examination. Much of my own enthusiasm has sprung from my admiration of the imaginative efforts of those who are taking the lead in this field.

These chapters have evolved from a course of unsophisticated lectures on "modern" mathematical topics. That they have finally reached their present form is due in no small measure to the generous help and encouragement of colleagues and friends.

In particular I wish to acknowledge with gratitude the ready response of two members of H.M. Inspectorate to my request for their personal reactions to the shape and content of the work, and to thank them for their generous encouragement and helpful advice.

I am greatly indebted to Professor R. L. Goodstein for permission to quote from certain mathematical notes published in recent issues of the *Mathematical Gazette*. I am also obliged to Random House for permission to quote two definitions from the book *Geometric Transformations* by I. M. Yaglom.

Finally, I would express my sincere thanks to the Pergamon Press; to the expert reader for his valuable criticisms and advice, to the editors who have assisted me during the various stages of preparation and to the compositors for the excellence of their work.

October, 1964

The third impression of this book incorporates a number of minor corrections, additions and alterations, some of which have been made following helpful suggestions gratefully received from readers of earlier impressions of the book.

June 1966   D .T. E. M.

For
J and J

CONTENTS

INTRODUCTION x

1 SETS 1
Idea of a set; notation; subsets; complement, union, intersection; Venn diagrams; sets of solutions; ordered pairs; relations and functions; mappings; laws of sets or the algebra of sets; applications

2 APPLICATIONS OF SET THEORY. BOOLEAN ALGEBRA 22
Premises, conclusions, Venn diagrams; statements, connectives, notation; alternative notation; Boolean algebra; simple applications to logic; simple applications to switching circuits

3 GROUPS 46
Binary operations; symmetry; the idea of a group; definition of a group; groups under addition; groups under multiplication; groups under other laws of combination; isomorphic groups; permutation groups

4 MATRICES 68
Rotation in the Euclidean and complex planes; quarter-turns in the Cartesian plane; general rotation in the Cartesian plane; enlargements; transformation of an area; general properties of matrices; symmetric and antisymmetric matrices; solution of linear

simultaneous equations using the inverse matrix; general method of forming the inverse matrix; an alternative method of determining the inverse matrix; matrix multiplication in arithmetical problems; eigenvectors and the characteristic equation of a matrix

5 VECTORS 99

Free vectors; representation and addition of vectors; position vectors and matrices; applications to geometry; resolution of a vector; the scalar product of two vectors; applications of the scalar product; the vector product of two vectors; vector area; applications of the vector product; applications to geometry; motion geometry

6 INEQUALITIES AND LINEAR PROGRAMMING 132

Lattices; open and closed half-planes; convex sets; linear programming; the transport problem; non-linear programming

7 STATISTICS 156

Collection of data; representation of data; frequency polygons and histograms; frequency curves and the normal distribution curve; mean mode, median; measures of dispersion; the semi-interquartile range; the mean deviation; the standard deviation; short method of calculating the standard deviation; the method of least squares—correlation; significance; the normal distribution curve

8 PATTERNS IN ARITHMETIC 184

The present situation; patterns in addition and multiplication; Chinese multiplication; numbers of the form $6N+1$; magic squares; magic matrices; latin squares; triangular numbers; square numbers; prime numbers; powers of numbers; fractional numbers and Farey series; the Fibonacci sequence; the golden section; indices and logarithms; binary slide rule; the binary scale and other scales

APPENDIX. EXERCISES ON CHAPTERS 1–7   219

HINTS AND SOLUTIONS TO EXERCISES   236

INDEX   249

INTRODUCTION

SOME years ago, on 29 September 1943, a conference was called on the initiative of the Cambridge Local Examinations Syndicate to consider an alternative syllabus in school mathematics. A rigid division into arithmetic, algebra and geometry at that time characterized school time-tables and the School Certificate examination papers of several examining Boards, and this was giving rise to signs of dissatisfaction. Moreover, the syllabus in geometry was extensive, formal and unsuitable for the weaker pupils. As a result a committee was set up under the chairmanship of Professor Jeffery and in 1944 produced what has become known as the "Jeffery Syllabus". Heavy manipulation in arithmetic and algebra was cut and formal geometry was reduced. The applications of geometry and trigonometry were merged and some elementary work in calculus was introduced. Papers set by the various Boards also contained elementary problems on latitude and longitude, plans and elevations and vector triangles of velocity.

We now find ourselves in a rather similar situation. Sixth-form pupils are leaving school after courses in advanced mathematics which have changed little in the last twenty years, only to find that courses in the universities have changed a great deal. Classical analysis is giving place to point set topology; geometry is developed out of the concept of vector spaces; group theory and numerical analysis now find a place in first year honours courses, and vector methods are employed freely in the teaching of dynamics and statics. Furthermore, those who proceed to technological studies encounter the need for a knowledge of matrix algebra, set theory, numerical analysis and vector algebra.

INTRODUCTION

The idea that algebra means "using letters instead of numbers" is out of date. Mathematicians now regard their subject as resting on a three-cornered foundation of algebra, topology and logic. Matrices, vectors and complex numbers should all be taught as different kinds of algebraical systems, and emphasis should be laid on the kinds of operations which can be performed and the laws to which they are subject.

This situation has claimed the attention of distinguished mathematicians throughout the world. In America an immense amount of curriculum research has been carried out, and thousands of children are working from experimental texts. The first of the "new mathematics" projects was the University of Illinois Curriculum Study in Mathematics which began in 1952. It is sponsored by the Carnegie Corporation of New York and the University of Illinois. The other large-scale project is based on Stanford University and is financed by the National Science Foundation. It publishes its texts through Yale University Press. In Europe the famous O.E.E.C. reports have been published and made widely and freely available.

In our own country the move towards reform gained momentum after the Southampton Conference of 1961 under the chairmanship of Professor Bryan Thwaites, which had been preceded by conferences in Oxford in 1957 and in Liverpool in 1959.

Out of the Southampton Conference emerged the first large-scale experiment in school mathematics teaching. The "School Mathematics Project" began with a group of eight schools and now embraces over forty. These schools have been working since 1962 with experimental texts, and a new O-level syllabus in "modern" mathematics has been accepted for the group by the Cambridge Local Examinations Syndicate, the Oxford and Cambridge Joint Board and the London University Examinations Board.

The "Midlands Mathematical Experiment", a project with similar aims, is in operation in the Birmingham area, and in addition many individual schools are seeking their own solution to the problem.

In general the changes involved are as follows:

1. There is an increase and change of emphasis in the algebraic content. Set notation is introduced both to add precision to the formulation of basic mathematical concepts and also as a means of unifying the various branches of the subject.

2. "Everyday mathematics" is exemplified by the inclusion of the elementary ideas of statistics, probability and linear programming. Stress is laid on intelligent approximation and the degree of accuracy appropriate to a given situation rather than to heavy routine computation. The use of slide rules and calculating machines, as a means both of saving labour and time and also of illustrating the mathematical ideas upon which these devices depend for their operation and instruction, is encouraged. Number concepts and patterns in the usual processes of elementary arithmetic become more vividly illustrated.

3. The various approaches to geometry present an interesting situation. In some schools which are carrying out experimental work the study of the Euclidean plane is by means of the geometrical transformations of translation, reflection and rotation. The properties of similar triangles and the ratio theorems are developed through the principle of enlargement. Treated in this way geometry ties up well with modern concepts of algebra. Using a letter to stand for an operation or transformation leads directly to the idea of a group. Meanwhile such transformations are also being achieved algebraically in the Cartesian plane by the use of matrices.

A parallel approach, however, is to develop elementary geometry through the use of vectors. Here there is another link with algebra (vector algebra), and in turn column vectors, which may be treated as matrices, lead to the idea of a group. Further, the use of vectors facilitates at a later stage a modern approach to the teaching of statics and dynamics. Vector geometry and motion geometry are discussed in greater detail in Chapter 5.

Changes of this nature give rise to a variety of reactions amongst those directly concerned. On the one hand, there is keen enthusiasm, on the other, entrenched opposition, but in between there is a great range of genuine concern and inquiry as to the exact nature of the proposed changes and their practical implications.

It is particularly for those who may feel that they come into this latter group that this book has been written.

The various chapters, with the possible exception of the one on statistics, cover only those topics which are completely new as far as the usual examination syllabuses are concerned. Even so, it is hoped that some degree of unity has been achieved. From the basic concept of a set and the algebra of sets is developed the idea of a Boolean algebra and its application to logic and switching circuits. In Chapter 3 we discuss those sets which, under some given law of combination, possess the properties of a group. Here is a broader approach to algebra and a link with geometry and the symmetries of geometrical figures. From groups of transformations in the Euclidean plane we proceed to those operators (matrices) which produce similar transformations in the Cartesian plane. Chapter 5 is concerned mainly with a vector approach to elementary geometry and with the connexion between matrix and vector algebra. The remaining chapters are concerned with rather more numerical aspects of mathematics and earlier concepts, notation and methods are used where appropriate. At a first reading Chapter 2 may be omitted, and Chapters 1, 3, 4, 5 and 6 might well be taken in that order. Chapter 8, which is intended to convey the desirability of looking for new patterns while using well-known processes, might equally well be taken first of all. Even though this chapter does contain some references to earlier chapters, its essential idea of taking a new look at already familiar topics underlies the whole book.

Thus the attempt has been made to provide, within the covers of one small book, a simple introduction to most of the topics which have found, or are likely in the next few years to find, their way into revised syllabuses in mathematics. It is hoped that this will be of use both to teachers who have hitherto had little or no opportunity of studying these topics, and also to students and sixth-form pupils who, having followed a more traditional course, are, nevertheless, likely to encounter the need for such knowledge in the future.

Each chapter concludes with a list of recently published texts suitable for further reading, while for those who wish to test their

INTRODUCTION

reading an appendix of elementary graded exercises and solutions is provided. Perhaps the best exercise of all for the reader is for him to construct a set of examples of his own.

Finally, for those who wish to read something of the present situation in school mathematics, the curriculum experiments and the history of the problem, the following publications are useful:

DIENES, Z. P., *Modern Mathematics for Children*, E.S.A.
DIENES, Z. P., *The Power of Mathematics*, Hutchinson.
FLETCHER, T. J., *Some Lessons in Mathematics*, C.U.P.
HERITAGE, R. S., *Learning Mathematics Book I* (The Shropshire Mathematics Experiment), Penguin.
HOPE, C., *A New O-level Syllabus in Mathematics*, Mathematics Teaching, No. 22.
HUGHES, M. G., *Modernizing School Mathematics*, Bell.
LAND, PROFESSOR F. W., *New Approaches to Mathematics Teaching*, Macmillan.
MOAKES, A. J., *The Core of Mathematics*, Macmillan.
SKEMP R. R., *The Psychology and Mathematics Project*, Mathematics Teaching No. 26.
SKEMP, R. R., *Understanding Mathematics*, Book I, London University Press.
THWAITES, PROFESSOR BRYAN, *On Teaching Mathematics*, Pergamon.
Director's Report, School Mathematics Project, University of Southampton.
Modern Mathematics for Schools Book I, Scottish Mathematics Group, Blackie and Chambers.
The Midlands Mathematical Experiment, Books I and II, Harrap.
School Mathematics Project, Book T and Book I, Southampton University, and C.U.P.
Synopses for Modern Secondary School Mathematics, O.E.C.D., 2 rue Andre Pascal, Paris 16.

1
SETS

THE idea of a set is a basic concept in "modern mathematics". At an advanced level classical analysis is gradually giving place to point set topology. The standard theorems remain but they assume a more general form. Group theory, with its important applications in the field of mathematical biology, is essentially the study of special types of sets of elements or operations. The modern definition of a geometry is a pair (S, G), where S is a set and G is a group of transformations on S. The language and notation of set theory and group theory are to be found in nearly every branch of modern mathematics at the university level. Moreover, Boolean algebra—and all such algebras are algebras of sets—has powerful practical applications in the spheres of symbolic logic and switching circuits.

Not only is the idea of a set a basic concept, it is also an extremely simple concept. Small children with their sets of bricks, sets of cards, sets of coloured beakers, etc. are well able to associate things of a kind, often before they can count. At the secondary stage the idea is capable of development along many fruitful lines as we shall see later. Indeed, if mathematics is to grow into an exciting world of patterns, relationships and structures instead of a succession of tricks and techniques, it must be child-centred and grow out of simple concepts such as this.

IDEA OF A SET

We do not define a set but we clarify our idea of a set to this extent: *a set is a collection of well-defined objects thought of as a whole*. For example, we may consider a set of spanners, a set of tools, a set of girls' names, the set of positive integers, a set of people or the set of ways in which we can arrange the digits 1, 2, 3. The objects, numbers, etc., comprising the set are generally called the *elements* of the set and they may be finite or infinite in number.

NOTATION

A set may be expressed either by *listing*, i.e. giving a list of its elements, or by giving a concise, unambiguous description of its elements.

Thus, for example, if A is the set of digits, J the set of integers, $J+$ the set of positive integers, we have

$$A = \{0, 1, 2, 3, 4, 5, 6, 7, 8, 9,\}$$

or $A = \{a|a \text{ is a digit}\}$ which reads "A is the set of elements a such that a is a digit".

$$J = \{\ldots -2, -1, 0, 1, 2, \ldots\} \quad \text{or} \quad \{n|n \text{ is an integer}\},$$
$$J+ = \{1, 2, 3, \ldots\} \quad \text{or} \quad \{p|p \text{ is a positive integer}\}.$$

In the case of A, "2 is an element of the set A", or, in standard notation, $2 \in A$. Similarly, $-1 \in J$, $3 \in J+$, etc. But $-2 \notin J+$ and $\cdot 5 \notin J$.

SUBSETS

A set will often contain a smaller set within itself, or we may say that a set contains a *subset* or subsets of itself. Thus a pack of cards or a set of cards contains, amongst others, the subset of clubs, the subset of picture cards and the subset of aces.

A universe $\mathscr{E}$ consisting of the three elements a, b, c, i.e.

$$\mathscr{E} = \{a, b, c\},$$

has the following *proper subsets*:

$$\{a, b\} \quad \{a, c\} \quad \{b, c\} \quad \{a\} \quad \{b\} \quad \{c\},$$

and in order to give complete generality to the idea of a subset we include the universal set $\{a, b, c\}$ itself and the *empty set* or *null set* $\{\ \}$ which contains none of the three elements. The empty set is usually regarded as a proper subset and is denoted by the symbol $\emptyset$.

Altogether the set $\{a, b, c\}$ has 8 subsets or 2^3 subsets. In general, *the set of n elements has 2^n subsets*. For taking each element in turn we may either include it or exclude it from a subset. To each element there corresponds 2 choices, thus to n elements there correspond $2 \times 2 \times 2 \ldots n$ times or 2^n choices. But each separate choice yields a different subset so that we have 2^n subsets altogether.

The sign $\subset$ means "is contained in" and $\subseteq$ means "is contained in or is equal to". If A, B, C are subsets of some universal set $\mathscr{E}$ we may write $\emptyset \subseteq A \subseteq \mathscr{E}$; but if A is a *proper subset* of $\mathscr{E}$ then $\emptyset \subseteq A \subset \mathscr{E}$.

If A and B are subsets of $\mathscr{E}$ and there is no element common to A and B, then the sets A and B are said to be *disjoint*. For example, if $\mathscr{E}$ is a pack or set of cards, the suit of spades is a proper subset and the suits of spades and clubs are disjoint subsets. However, the set of club cards and the set of deuces are not disjoint. We say that they *intersect* and that their intersection contains the single element the deuce of clubs.

COMPLEMENT, UNION, INTERSECTION

The *complement* of a set A is denoted by $\mathscr{C}A$ or by A' and is the set of those and only those elements which are *not* in A.

The *union* $A \cup B$ (pronounced A union B or A cup B) of A and B is the set of those and only those elements which are in A or in B (or both).

The *intersection* $A \cap B$ (pronounced A intersection B or A cap B) of A and B is the set of those and only those elements which are in A and in B.

For example: consider the universal set

$$\mathscr{E} = \{\text{Alf, Bert, Claud, Dave, Eric, Fred}\}.$$

This has subsets

$$A = \{\text{Alf, Bert, Claud}\},$$
$$B = \{\text{Bert, Claud, Dave, Eric}\},$$
$$C = \{\text{Alf, Fred}\} \quad \text{(as well as 61 others)}.$$

In this case

$$A' = \{\text{Dave, Eric, Fred}\},$$
$$B' = \{\text{Alf, Fred}\} = C.$$
$$A \cup B = \{\text{Alf, Bert, Claud, Dave, Eric}\}$$
$$\text{(list repeated elements once only)},$$
$$A \cap B = \{\text{Bert, Claud}\},$$
$$B \cup C = \mathscr{E},$$
$$B \cap C = \emptyset.$$

Further $(A \cup B)' = \{\text{Fred}\} = A' \cap B'$

and $(A \cap B)' = \{\text{Alf, Dave, Eric, Fred}\} = A' \cup B'$

the last two of which are important general results in the algebra of sets incorporating a dual property.

VENN DIAGRAMS

At an advanced level a diagram forms no part of a rigorous proof, and such proofs in set algebra are conducted in terms of elements and the operations of inclusion, complementation, union and intersection. At an introductory or stage A level, however, the whole subject is illuminated and clarified by the use of Venn diagrams. At this stage we are not concerned with rigour; on the contrary, uninhibited experiment and imaginative investigation with such diagrams can lead a child to discover the majority of the basic relationships in set algebra for himself. Here the child can build structures for himself, and surely it is this type of "research" at the embryonic stage which we must encourage whenever possible if we are to produce creative thinkers and mathematicians.

Usually we represent the universal set $\mathscr{E}$ by a rectangular area (Fig. 1) and its subsets by circles within this area.

If A is a subset of $\mathscr{E}$ then the area inside $\mathscr{E}$ but outside A represents the set A'.

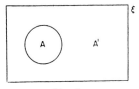
Fig. 1

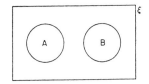
Fig. 2

If A and B are disjoint the circular areas do not overlap (Fig. 2). Whereas if A and B are not disjoint their union and intersection are represented by the shaded areas shown in Figs. 3 and 4.

We have now covered all the notation and most of the vocabulary necessary for an O-level or pre-sixth-form course. It will have

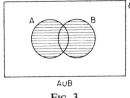
A∪B
Fig. 3

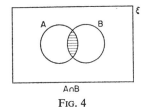
A∩B
Fig. 4

been presented far more gradually than in this chapter, and at each stage the ideas and words will be made familiar through numerous examples and exercises, oral and written. Indeed, much of the development will arise out of the exercises themselves. The lines of development from this point onwards may well proceed by giving many examples of this type.

Write down the set of all months in the year with initial letter J

$\{\text{January, June, July}\} = J$, say.

Write down the set of all months having 30 days

$\{\text{April, June, September, November}\} = S$, say.

If $\mathscr{E}$ is {January, February, March, ..., December} write down

(a) by inspection,
(b) using a Venn diagram,

the sets J', S', $J \cup S$, $S \cap J$, $J \cap S'$, $\mathscr{E} \cap J$, $\mathscr{E} \cup J$, etc.

Two sets are equal if they each contain the same elements.

Example

Using the signs $=$, $\neq$, $\subset$ what statements can you make about the following pairs of sets?

$$A = \{1, 4, 7, 10, 13\} \quad B = \{4, 7, 10, 13, 16\}$$
$$C = \{p, q, r\} \quad D = \{p, q\}$$
$$E = \{\triangle \square \bigcirc\} \quad F = \{\bigcirc \triangle \square\}$$
$$G = \{\bigcirc\} \quad H = \emptyset$$

Here $A \neq B$, $D \subset C$, $E = F$, $H \subset G$ (for $\bigcirc$ is an element and $\emptyset$ contains no elements).

SETS OF SOLUTIONS

Given some universal set and some restriction upon this set, then the elements which satisfy this restriction form the *set of solutions*. The restriction may take the form of a simple statement or condition, an equation or an inequality and, of course, at this point there occurs the opportunity to examine and use the inequality signs $>$ and $<$. Just as we express relationships between sets in the form $A \subset B$, so we express relationships between elements, or quantities or numbers in the form $a < b$ where this is possible.

Example

If $\mathscr{E} = \{1, 2, 3, 4, 5, 6, 7, 8, 9\}$ list the subset of $\mathscr{E}$ which satisfies, or the set of solutions of the following:

(i) x is an even number,   (iv) $3x + 1 = 10$,
(ii) $5 + x > 10$,   (v) $x + 4 = 15$,
(iii) $5 + x < 10$,

and the required subsets are

(i) $\{2, 4, 6, 8\}$, (iv) $\{3\}$,
(ii) $\{6, 7, 8, 9\}$, (v) $\emptyset$.
(iii) $\{1, 2, 3, 4\}$,

Example

If $\mathscr{E}$ is the set of all the natural numbers list the set of solutions of

$\{x | 3x-1 = 5\}$ i.e. $\{2\}$
$\{x | x+2 > 5\}$ $\{4, 5, 6, \ldots\}$
$\{x | x+6 = 6+x\}$ $\mathscr{E}$
$\{x | \frac{1}{2}x = 5\}$ $\{10\}$
$\{x | 5x-1 = 2\}$ $\emptyset$

This sort of work will lead to more examples on equalities. Emphasis in the past has been on the formation and solution of equations. To the child, however, the concept of equality is no more simple than the concept of inequality. In real life, ages, weights and quantities are more often unequal than otherwise, in fact equality is a comparatively rare phenomenon. Variety is the spice of life, so it is said, while equality of wealth and opportunity is an almost unattainable ideal in the view of others. "My brother is bigger than yours", "Our team is better than theirs", are common childish boasts; how often do we hear comments of this type expressing equality? If mathematics is to arise out of experience, then surely the case of equality is a special case of a general concept of inequality and should be treated as such. It is true that many relationships or patterns are beautiful simply because the ideal state of equality exists in some respect or other. If the three angles of a triangle are equal each to each the triangle possesses a beauty of symmetry and a number of interesting properties. However, it is not the business of the mathematician to study equalities exclusively, and to this extent we should and could do more work with inequalities.

ORDERED PAIRS

Let X and Y be two sets and let x, y be typical elements of each set respectively. Then $x \in X$ and $y \in Y$. The set (x, y) is then called an *ordered pair*. It is obviously a pair and it is ordered in the sense that x (from X) is always taken first.

To illustrate in diagrammatic form (Fig. 5), the elements of X can be marked off on a horizontal line (the X-axis) and those of Y on a vertical line (the Y-axis). If $X = \{1, 2\}$, $Y = \{1, 2, 3\}$, then

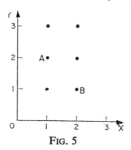

Fig. 5

the set of all ordered pairs (x, y) is represented as shown in the figure. Clearly the ordered pair $(1, 2)$ (A) is not the same as the ordered pair $(2, 1)$ (B).

If $X = \{x | x \text{ is an integer}\}$, $Y = \{y | y \text{ is an integer}\}$, the sets of ordered pairs form an infinite square *lattice* of points.

The next step is the representation diagrammatically of the sets of ordered pairs $\{(x, y) | ax + by = c\}$ and $\{(x, y) | ax + by > c\}$. This work is done most effectively in the classroom by providing each child with a rectangular piece of pegboard and a set of coloured plastic pegs. The apparatus can be used in connexion with linear laws, intersections of sets of ordered pairs, the solution of simultaneous linear equations, simple problems in linear programming, the ideas of a relation, a function and a mapping.

Example

Illustrate diagrammatically the set of ordered pairs

$$\{(x, y) | x + y = 10\} \qquad x \in J, y \in J.$$

A listing would contain $\{\ldots (0,10)(1,9)(2,8)\ldots\}$ represented as shown in Fig. 6.

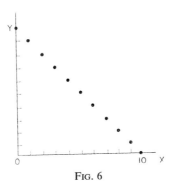

Fig. 6

Example

Illustrate graphically the set of ordered pairs

$$\{(x,y) | y > x-2\}$$

Add pegs or points to give

$$\{(x,y) | y \geqq x-2\}$$

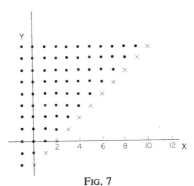

Fig. 7

The added points then appear as crosses in Fig. 7.

RELATIONS AND FUNCTIONS

[This section may be omitted at a first reading without loss of continuity]

The set of all ordered pairs (x,y) where $x \in X$, $y \in Y$ is called the *product set* of X and Y and is denoted $X.Y$.

Any proper subset of $X.Y$ is called a *relation* between the elements of X and those of Y. Thus, for example, if

$$X = \{1,2\}, \qquad Y = \{1,2,3\},$$

then $\qquad \{(1,1)(1,2)\} \quad$ or $\quad \{(1,1)(2,1)(2,3)\}$
or $\qquad \{(1,1)(2,2)\} \quad$ or $\quad \{(1,1)(2,3)\}$

are *relations*.

If $\quad X = \{\text{John, Jack}\}, \qquad Y = \{\text{Sheffield, London}\}$,
then $\qquad \{(\text{John, Sheffield})(\text{John, London})\}$
and $\qquad \{(\text{John, Sheffield})(\text{Jack, London})\}$

are relations.

A relation is a *function* if and only if no two different ordered pairs of the relation have the same first member. Thus in the cases for $X = \{1,2\}$, $Y = \{1,2,3\}$ above, $\{(1,1)(2,2)\}$ and $\{(1,1)(2,3)\}$ are *functions*. If X and Y comprise the set of all real numbers, then $\{(1,1)(2,2)\}$ might lie on the line $y = x$ (identity) and $\{(1,1)(2,3)\}$ might lie on the line $y = 2x-1$. Normally we know these as linear functions. Note, however, $\{(1,1)(2,2)\}$ could lie on $y = x^2 - 2x + 2$ or any number of other familiar functional relationships between x and y. Again, in the second example, $\{(\text{John, Sheffield})(\text{Jack, London})\}$ is a function but $\{(\text{John, Sheffield})(\text{John, London})\}$ is not a function.

By these definitions in terms of sets we see that we include not only all our traditional ideas of a function but many other ideas as well. Our idea of a function becomes broader and far more general. At the same time the idea is extended backward into far simpler types of relationship which hitherto we should not have regarded as "functions" at all.

MAPPINGS

A mapping is a function; no essentially new idea is involved. However, as the term implies, it is a more diagrammatic and often helpful way of looking at a function. {(John, Sheffield)(Jack, London)} is a function; it is also a mapping, Fig. 8, or to children,

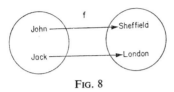

Fig. 8

Fig. 9. Here we may think of f as a translation. Generally f can be any form of relation which obeys the rules given above for a function. It may involve many forms of *transformation*, e.g. a translation, a rotation, a purely algebraic substitution or simply a 1:1 or many-one correspondence between two sets of elements.

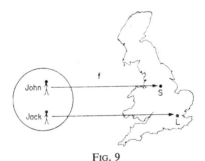

Fig. 9

In this case it might be that John and Jack return to their respective homes Sheffield and London at the end of their boarding-school term and f sets up a 1:1 *correspondence* between each boy and his home town. Functions do not always involve a 1:1 correspondence between the elements of two sets. In fact this is a special case; the more general case is that of a many-one correspondence. For example, suppose $X = $ {John, Jack, David} and $Y = $ {Sheffield,

London} and that Jack and David both live in London. We then have Fig. 10. The ordered pairs of this relation are (John, Sheffield) (Jack, London)(David, London), and as no two ordered pairs have the same first element the relation is still a function.

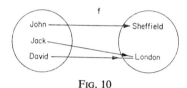

Fig. 10

However, John cannot live both in Sheffield and in London. So although the ordered pairs of $X.Y$ include {(John, Sheffield) (John, London)}, this set would not define a function, for here we have two different ordered pairs with the same first element. Given the sets $X = $ {John, Jack, David} and $Y = $ {Sheffield,

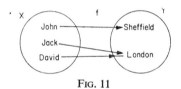

Fig. 11

London} and the function (Fig. 11), we introduce the notation for a function

$$X \xrightarrow{f} Y$$

If x, y are typical elements respectively of the sets X, Y we may also write $y = f(x)$. The set of elements x in X for which $f(x)$ exists in Y is sometimes called the DOMAIN over which the function f operates. The corresponding values of $f(x)$ in Y for which x exists in X is called the RANGE.

As a numerical example we might take X as the set of real numbers and Y as the set of real numbers such that $y = x^2$. The ordered pairs of $X.Y$ include, for example, $(1,1)$, $(2,4)$, $(2\frac{1}{2}, 6\frac{1}{4})$, $(-\frac{1}{2}, \frac{1}{4})$, etc. There are no ordered pairs which are different and yet which have the same first element. Thus $y = x^2$ is a functional relationship or $f(x) = x^2$. If, however, $y^2 = x$, we do not have a functional relationship, for $X.Y$ would then contain the ordered pairs $(1, 1)$ and $(1, -1)$. These are different ordered pairs with the same first element and hence they do not satisfy the criterion of a function.

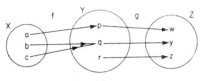

Fig. 12

Problems sometimes involve the successive operation of two or more functions (Fig. 12).

Here $$X \xrightarrow{f} Y \quad \text{and} \quad Y \xrightarrow{g} Z$$
or $$f: X \to Y \quad \text{and} \quad g: Y \to Z$$

And, for example

$$f(a) = p \quad \text{and} \quad g(p) = w$$

so that we write

$$g_0 f(a) = g(p) = w$$

(some writers put $g(f) a$ or $gf(a) = w$) and members of the set X are mapped on to members of the set Z under the transformation $g_0 f$.

As an example of a product of mappings we have

$$x \to x+2; \quad x \to x^3; \quad x \to \tan x$$

yields the composite function $\tan (x+2)^3$, or, written in another way

$$f(x) = x+2, \quad g(x) = x^3, \quad h(x) = \tan x$$
then $$h_0 g_0 f(x) = \tan (x+2)^3$$
or $$hgf(x) = \tan (x+2)^3$$

In the case (Fig. 13) $f: X \to Y$ is a one-to-one mapping of X on to Y. If $g: Y \to X$ such that $gf: X \to X$ is the identity, then g is called the *inverse* of f and we write $f^{-1}: Y \to X$. The idea of a 1:1 correspondence is of great importance in mathematics and is the basic concept in counting. Children will be able to think of many examples involving a one-to-one correspondence between two

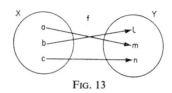

FIG. 13

sets. Cloakroom peg numbers and children's names, Morse code symbols and letters of the alphabet, etc., may be discussed from this point of view. If more than one child uses one peg we have a many-one mapping. If each child has his own peg and every one is used we have a 1:1 mapping of one set *onto* another; if each child has his own peg and not all pegs are required, we have a 1:1 mapping of one set *into* another and so on.

In the example on unit square lattices we have an opportunity to discuss at a more advanced level the elements of $\mathscr{E}$ such that $\{(x,y) | ax+by > c\}$ is an area instead of a lattice of isolated points. Will this be the case if $\mathscr{E}$ = set of all the rationals? After all, between any two fractions $\dfrac{a}{b}$ and $\dfrac{c}{d}$ there always lies a third $\dfrac{a+c}{b+d}$. This might form as good a starting point as any for a consideration of the irrational numbers.

LAWS OF SETS OR THE ALGEBRA OF SETS

By investigation or well-directed exercises on Venn diagrams the child will be able to discover most of the following for himself:

$$\left. \begin{array}{l} A \cup B = B \cup A \\ A \cap B = B \cap A \end{array} \right\} \text{the commutative law}$$

$$\left. \begin{array}{l} (A \cup B) \cup C = A \cup (B \cup C) \\ (A \cap B) \cap C = A \cap (B \cap C) \end{array} \right\} \text{the associative law}$$

together with

1. $A \cap \mathscr{E} = A$
2. $A \cup \emptyset = A$
3. $A \cap (B \cup C) = (A \cap B) \cup (A \cap C)$
4. $A \cup (B \cap C) = (A \cup B) \cap (A \cup C)$
5. $A \cap (A \cup B) = A$
6. $A \cup (A \cap B) = A$
7. $A \cap \emptyset = \emptyset$
8. $A \cup \mathscr{E} = \mathscr{E}$
9. $A \cap A = A$
10. $A \cup A = A$
11. $(A \cup B)' = A' \cap B'$
12. $(A \cap B)' = A' \cup B'$

We shall leave the verification of most of these to the reader, but we demonstrate below a diagrammatic "proof" of 3, 4 and 11.

3. The horizontally shaded area (Fig. 14) is $B \cup C$ and the

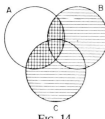

Fig. 14   Fig. 15

cross-hatched area is $A \cap (B \cup C)$. The vertically shaded area (Fig. 15) shows both $(A \cap B)$ and $(A \cap C)$. Thus the shaded area is the sum of these areas, i.e. the total shaded area is

$$(A \cap B) \cup (A \cap C)$$

But the shaded area here and the cross-hatched area in Fig. 14 are equal so that

$$A \cap (B \cup C) = (A \cap B) \cup (A \cap C), \quad \text{which is 3.}$$

4. Vertical shading (Fig. 16) gives $(B \cap C)$ while the total shaded

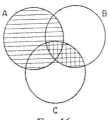

Fig. 16

area is $A \cup (B \cap C)$. The horizontally shaded area (Fig. 17) shows $(A \cup C)$. The vertically shaded area is $(A \cup B)$. These areas intersect in the cross-hatched area $(A \cup B) \cap (A \cup C)$. But the

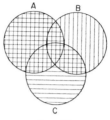

Fig. 17

shaded area in Fig. 16 is the same as the cross-hatched area in Fig. 17, so that

$$A \cup (B \cap C) = (A \cup B) \cap (A \cup C), \quad \text{which is 4.}$$

11. The shaded area (Fig. 18) is $(A \cup B)$. Thus the unshaded area in $\mathscr{E}$ is $(A \cup B)'$. A' is shaded horizontally and B' is shaded vertically. The cross-hatched area (Fig. 19) is thus the intersection

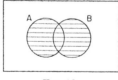

Fig. 18

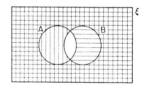

Fig. 19

of A' and B' or $A' \cap B'$. But the unshaded and cross-hatched areas are equal so that

$$(A \cup B)' = A' \cap B', \quad \text{which is 11.}$$

All the other results may be demonstrated in the same way. Note particularly the *dual property* whereby unions and intersections are interchangeable in 3 and 4, 5 and 6, 9 and 10, and 11 and 12.

These diagrammatic methods of demonstrating the laws of sets, while sufficient in the early stages, are not proofs. At the sixth-form level much more rigour is desirable and as examples of this proofs of case 4 and case 11 are given below.

Case 4. $\qquad A \cup (B \cap C) = (A \cup B) \cap (A \cup C).$

Let $\qquad x \in A \cup (B \cap C)$
then either $\qquad x \in A$ or $x \in (B \cap C).$
If $x \in A$, then $\qquad x \in A \cup B$ and $x \in A \cup C,$
i.e. $\qquad x \in (A \cup B) \cap (A \cup C).$
If, however, $\qquad x \in (B \cap C),$ then $x \in B$ and $x \in C.$
Hence $\qquad x \in A \cup B$ and $x \in A \cup C,$
i.e. $\qquad x \in (A \cup B) \cap (A \cup C).$
We have shown that if $x \in A \cup (B \cap C)$, then
$$x \in (A \cup B) \cap (A \cup C).$$

It follows that

$$A \cup (B \cap C) \subset (A \cup B) \cap (A \cup C). \qquad (1)$$

Suppose now that $x \in (A \cup B) \cap (A \cup C),$
then $\qquad x \in A \cup B$ and $x \in A \cup C,$
i.e. either $\qquad x \in A,$ or $x \in B$ and $x \in C.$
Therefore either $\qquad x \in A$ or $x \in B \cap C.$
Hence $\qquad x \in A \cup (B \cap C)$
and we have

$$(A \cup B) \cap (A \cup C) \subset A \cup (B \cap C). \qquad (2)$$

From (1) and (2) it follows that

$$A \cup (B \cap C) = (A \cup B) \cap (A \cup C).$$

Case 11. $\qquad (A \cup B)' = A' \cap B'.$

Suppose that $\qquad x \in (A \cup B)',$
then $\qquad x \notin A \cup B,$
i.e. $\qquad x \notin A$ and $x \notin B.$
Hence $\qquad x \in A'$ and $x \in B'.$
$\qquad \therefore x \in A' \cap B',$

whence $\qquad (A \cup B)' \subset A' \cap B'. \qquad (1)$

Suppose now that $x \in A' \cap B'$,
then $\quad x \in A'$ and $x \in B'$
or $\quad x \notin A$ and $x \notin B$,
i.e. $\quad x \notin (A \cup B)$.
Hence $\quad x \in (A \cup B)'$.

$$\therefore A' \cap B' \subset (A \cup B)'. \tag{2}$$

From (1) and (2) it follows that

$$(A \cup B)' = A' \cap B'.$$

APPLICATIONS

Venn diagrams are useful when we require to analyse statistical information involving overlapping classes.

Example

A certain science sixth form contains 50 pupils all of whom take mathematics. 18 study chemistry, 17 study biology, 24 study physics. Of those taking three subjects, 5 study physics and chemistry, 7 study physics and biology, and 6 study chemistry and biology while 2 take all four subjects. How many pupils study only mathematics?

Here let the universe be the set of those taking mathematics M with intersecting subsets physics P, chemistry C, and biology B. If we start by filling in the numbers in the intersections of these subsets starting with $n(M \cap P \cap C \cap B) = 2$ we have the Venn diagram (Fig. 20).

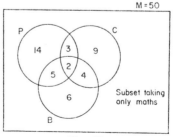

Fig. 20

Now $n(P \cup C \cup B) = 43$.

Therefore the number who study only mathematics

$$= n(P \cup C \cup B)' = 50 - 43 = 7.$$

Example

At a certain conference of 100 people there are 29 English women and 23 English men. Of these English people 4 are doctors and 24 are either men or doctors. There are no foreign doctors. How many women doctors are attending the conference?

Here we take a universal set of people P with a subset E (English people) having two intersecting subsets of its own; M (men) and D (doctors). If there are x lady doctors at the conference the Venn diagram appears as Fig. 21.

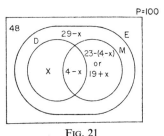

Fig. 21

Since 24 are either men or doctors,

therefore $\qquad x + 4 - x + 19 + x = 24.$

Hence $x = 1$, i.e. there is one woman doctor present.

Example

An analysis of 100 personal injury claims made upon a motor insurance company revealed that loss or injury in respect of an eye, an arm or a leg occurred in 30, 50 and 70 cases respectively. Claims involving the loss or injury to two of these members numbered 44. How many claims involved loss or injury to all three? We must assume that one or other of the three members was mentioned in each of the 100 claims.

Let x be the number required. Label the other intersections p, q and r. Let the sets of cases involving eyes, arms and legs be denoted by E, A and L respectively. From this data we produce the Venn diagram (Fig. 22).

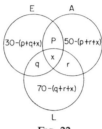

Fig. 22

The set $E \cup A \cup L$ contains 100 elements,

i.e. $30-(p+q+x)+50-(p+x+r)+70-(q+x+r)$
$\qquad +p+q+r+x = 100,$
i.e. $\qquad p+q+r+2x = 50,$
but $\qquad p+q+r = 44,$
therefore $\qquad 2x = 6,$

i.e. the number of claims involving all three members numbers 3.

It is unlikely that the development of set theory or the difficulty of exercises on it will go much beyond this point at the O-level or pre-sixth-form stage. In the sixth form, however, the idea of sets may be extended to Boolean algebra with its applications to symbolic logic and the theory of switching circuits. In the next chapter we discuss these matters.

Useful Reference Books

ADLER, I., *The New Mathematics*, New American Library.
AIKEN and BESEMAN, *Modern Mathematics—Topics and Problems*, McGraw-Hill, New York.
ALLEN, R. G. D., *Basic Mathematics*, Macmillan.
ALLENDOERFER and OAKLEY, *Principles of Mathematics*, McGraw-Hill, New York.
BAILEY, C. A. R., *Sets and Logic I and II*, E. Arnold.
BALFOUR, A., *An Introduction to Sets, Groups and Matrices*, Heinemann.

BOWRAN, A. P., *Sets for Schools*, Macmillan.
FLETCHER, T. J., *Some Lessons in Mathematics*, C.U.P.
GOODSTEIN, R. L., *Fundamental Concepts in Mathematics*, Pergamon Press.
HARRISON, P. G., *On Introducing Sets to Children*, Mathematics Teaching Pamphlet No. 10.
HART, SCHULT and BRISTOL, *Introduction to Sets and Inequalities*, D. C. Heath.
KEMENY, J. G., MIRKIL, H., SNELL, J. L. and THOMPSON, G. L. *Finite Mathematical Structures*, Prentice-Hall.
KEMENY, J. G., SNELL, J. L. and THOMPSON, G. L., *Introduction to Finite Mathematics*, Prentice-Hall.
KURATOWSKI, K., *Introduction to Set Theory and Topology*, Pergamon Press.
LAND, F., *The Language of Mathematics*, Murray.
LANDIN, H., *Set Theory*, Prentice-Hall.
MCKLEAN, K. R., *The Teaching of Sets in Schools*, Bell and Sons Ltd.
MOAKES, A. J., *The Core of Mathematics*, Macmillan.
SAWYER, W. W., *Prelude to Mathematics*, Freeman.
SCHAAF, W. L., *Basic Concepts of Elementary Mathematics*, Wiley.
SMITHERS, G., Modern mathematics in fifth forms, *Mathematical Gazette*, Vol. XLVII, No. 359 (Feb. 1963), Bell.
SELBY and SWEET, *Sets, Relations, Functions: An Introduction*, McGraw-Hill, New York.
TOSKEY, B., *College Algebra*, Addison-Wesley.
WHITESITT, E., *Boolean Algebra and its Applications*, Addison-Wesley.
Insights in Modern Mathematics. 23rd Year Book, National Council of Teachers of Mathematics, Washington D.C.

2
APPLICATIONS OF SET THEORY.
BOOLEAN ALGEBRA

(This chapter may be omitted at a first reading without loss of continuity)

PREMISES, CONCLUSIONS, VENN DIAGRAMS

Consider the argument:

Common salt is a chloride,	(1)
All chlorides are compounds,	(2)
Therefore common salt is a compound.	(3)

The statements labelled (1) and (2) are usually described as *premises* or *hypotheses*, and (3) is the *conclusion*. In this case the argument is *valid* and we say that the whole statement comprising (1), (2) and (3) has *truth value T*.

FIG. 23

We can demonstrate the validity of this argument by means of the simple Venn diagram shown in Fig. 23, where $\mathscr{E}$ is the set of all compounds, C is the set of all chlorides and s is the element†

† "Element" is, of course, to be taken in the mathematical and not the chemical sense.

common salt. Or we may dispense with the diagram and argue that

from (1) $s \in C$
and from (2) $C \subset \mathscr{E}$.
Hence from (1) and (2) $s \in \mathscr{E}$, which is 3.

In this case the conclusion not only follows from the argument but is, in fact, a true statement. However, from our point of view the truth or otherwise of a conclusion is irrelevant to the question as to whether an argument is valid. Thus the argument:

Paris is in Ohio,
Ohio is in America,
Therefore Paris is in America,

is a valid argument. The conclusion is incorrect because the first premise is untrue.

This sort of argument seems trivial and useless, but we do in fact sometimes use arguments of this sort in mathematics. They are referred to as *reductio ad absurdum* arguments. For a valid argument in which the conclusion is absurd reveals falsehood in one or other of the premises.

As an example, if we wish to prove that the tangent to a circle is perpendicular to the radius drawn to its point of contact, we may take as a premise the statement that the radius and tangent are *not* perpendicular. We then pursue a valid argument leading to, say, the conclusion that the tangent cuts the circle in two points, an obvious contradiction with the definition of the tangent to a circle, and hence we reveal the error in the premise. The device is also commonly used in analysis. To prove the irrationality of $\sqrt{2}$ we show that the premise $\sqrt{2} = \dfrac{a}{b}$, where a and b are coprime integers, produces a contradiction.

It is said that "two wrongs do not make a right", but in mathematics this is not always so. Two false premises can be used to produce a conclusion which is true by an argument which is quite valid. Some schoolboys are, in fact, quite adept at this (particularly if the answers are given at the back of the book). For example the premises $4 > 7$, $7 = 3$ lead to the true conclusion that $4 > 3$. Again:

All criminals are men,
All policemen are criminals,

leads by a valid argument to the true conclusion that all policemen are men (Fig. 24). For $P \subset C \subset M$, thus $P \subset M$.

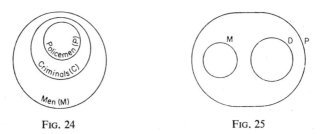

Fig. 24   Fig. 25

Venn diagrams and set algebra are often useful in checking the validity of an argument. Consider the argument:

Some problems are mathematical,
Some problems are difficult,
∴ some mathematical problems are difficult.

The conclusion is certainly true (at any rate in the experience of the author), and at first sight may seem to follow from the premises which are equally true. This, however, is not the case. In fact the argument is invalid. For if P is the set of all problems, M the set of all mathematical problems, and D the set of all difficult problems, we see from the Venn diagram (Fig. 25) that M and D do not *necessarily* intersect, i.e. $M \cap D$ can be the empty set $\emptyset$ which contradicts the conclusion that some mathematical problems are difficult problems, or that some

APPLICATIONS OF SET THEORY

elements of M are also elements of D, or that $M \cap D \neq \emptyset$. Treating the problem algebraically we have

$$P \cup M = P$$
$$P \cup D = P,$$
$$\therefore (P \cup M) \cap (P \cup D) = P \cap P = P.$$

But, by (4) of the laws of sets in Chapter 1,

$$(P \cup M) \cap (P \cup D) = P \cup (M \cap D).$$
Thus $\qquad P \cup (M \cap D) = P.$
Now $\qquad P \cup \emptyset = P \quad [\text{law (2)}].$
$\therefore M \cap D$ can have the value $\emptyset$.

As another example consider the argument:

All physicists are scientists,
No scientists are artists,
$\therefore$ no physicists are artists.

This is quite valid though the conclusion is questionable. Denoting the various sets as physicists P, scientists S, artists A, we

Fig. 26

have the Venn diagram (Fig. 26), and clearly P and A are disjoint. Alternatively the premises reduce to

$$P \cap S = P$$
and $\qquad S \cap A = \emptyset.$
Now $\qquad P \cap A = (P \cap S) \cap A$
$\qquad\qquad\quad = P \cap (S \cap A) \quad \text{by the associative law of sets}$
$\qquad\qquad\quad = P \cap \emptyset$
$\qquad\qquad\quad = \emptyset.$

$\therefore P$ and A are disjoint, therefore no physicists are artists.

STATEMENTS, CONNECTIVES, NOTATION

It will be clear to the reader that in the previous examples the people or objects involved are irrelevant to the structure of the argument. The argument:

All teachers are strict,
No strict people are fat,
∴ no teachers are fat,

is essentially the same argument as the one concerning physicists, scientists and artists. We may, therefore, use small letters to represent statements or premises, and if we do so there is clearly a very close connexion between the algebra by which we must manipulate these letters and the algebra of sets. Corresponding to the statements:

Jack is a cricketer c,
Jack is a tennis player t,

we have the set of all cricketers C, and the set of all tennis players T. Some writers describe T as the *truth set* of the statement t.

The set $C \cup T$ clearly contains all people who are *either* cricketers *or* tennis players; $C \cap T$ contains only those people who are both cricketers *and* tennis players.

We now introduce a notation which is used in the algebra of propositions to connect simple statements.

For two assertions or statements a and b, the *negation* of a is represented by $\sim a$ (*not a*); the assertion (disjunction) of either *a or b* (or both) is represented by $a \vee b$; the assertion (conjunction) of *a and b* is represented by $a \wedge b$.

The assertions of implication and equivalence are:

$a \rightarrow b$ or "a implies b",
$a \leftrightarrow b$ or "a implies b and vice versa",
    or "a if and only if b".

Thus $c \vee t$ means "Jack is a cricketer *or* he plays tennis", and $c \wedge t$ means "Jack is a cricketer *and* a tennis player". $\sim c \vee t$ means "either Jack is *not a cricketer or* he is a tennis player".

APPLICATIONS OF SET THEORY

If we take the statements:

The man is a soldier a,
The man is a policeman b,
The man is entitled to wear uniform c,

these may be connected thus: $a \vee b \to c$. For the man may wear uniform either if he is a soldier or a policeman (or both). But we may *not* write $a \vee b \leftrightarrow c$, for the fact that a man is entitled to wear uniform does not necessarily imply that he is a soldier or a policeman; he may be a sailor or a scoutmaster.

On the other hand, for example:

He is a minor p,
He is British q,
He is entitled to vote in the British General Election r.

Neglecting for a moment laws governing the franchise of peers, convicts and insane persons, we may connect these statements in the form:

If he is British and over 21 then he is entitled to vote; further, if he is entitled to vote in the British General Election then he must be British and over 21, or

$$\sim p \wedge q \leftrightarrow r.$$

To take a different example from switching circuits, consider the circuits shown in Fig. 27 (i) and (ii).

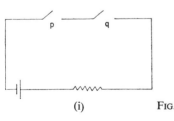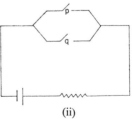

Fig. 27

Suppose we have statements:

The switch p is closed p,
The switch q is closed q,
Current flows in the circuit c.

Then in (i) we have $p \wedge q \leftrightarrow c$, for if both switches are on the current flows. Conversely, if the current is flowing then both switches must be closed.

Likewise in (ii) $p \vee q \leftrightarrow c$.

These examples serve to show roughly how two statements may be combined and how the connexion between them may imply, or be equivalent to, some third assertion c.

It is necessary, however, to treat these matters rather more precisely. The simplest assertion is that of negation. Clearly if p is true (T) then $\sim p$ is false (F) and vice versa. If we tabulate these values we have a truth table in its simplest form.

p	$\sim p$
T	F
F	T

In the case $p \wedge q$, four possibilities exist, p and q may be either (1) both true, (2) both false, (3) p true and q false, (4) p false and q true. Clearly the compound statement p and q can only be true if both p and q are true. The truth table is shown as Table 1.

TABLE 1

p	q	$p \wedge q$
T	T	T
T	F	F
F	T	F
F	F	F

Now consider the disjunction $p \vee q$. If one statement is true and the other false the disjunction is true.

Spanners are tools *or* All triangles are congruent

is clearly a true statement. If both are false the disjunction is clearly false. We cannot possibly say that

All dogs are mad *or* All men are sane.

When both statements are true, however, a difficulty arises. If I am wearing brown shoes and I am wearing a green hat is it true to say "I am wearing brown shoes *or* a green hat"? The answer clearly depends upon whether the disjunction $\vee$ in $p \vee q$ is taken to mean *p or q or both* (the inclusive disjunction), *or p or q but not both* (the exclusive disjunction). In practice both are used, the exclusive disjunction being distinguished by the symbol $\underline{\vee}$. In this chapter, however, as in the majority of cases, we shall take $\vee$ to stand for the inclusive disjunction, and in this case it is clearly permissible for me to state that I am wearing brown shoes or a green hat (or both). The truth table is shown as Table 2.

TABLE 2

p	q	$p \vee q$
T	T	T
T	F	T
F	T	T
F	F	F

More complicated expressions can now be built up. As an example consider the compound statement $(p \vee q) \vee \sim p$. Starting with all possible values of p and q we evaluate first $p \vee q$, then $\sim p$ and finally $(p \vee q) \vee \sim p$. Thus in Table 3

TABLE 3

p	q	$p \vee q$	$\sim p$	$(p \vee q) \vee \sim p$
T	T	T	F	T
T	F	T	F	T
F	T	T	T	T
F	F	F	T	T

$(p \vee q) \vee \sim p$ is an example of a statement which is *logically true*, i.e. is a correct statement under all possible circumstances. Referring to our previous example, this result tells us that "Either I am wearing brown shoes or a green hat (or both), or I am not wearing brown shoes" is a true statement no matter what I am wearing.

This fact is not immediately obvious from the statement itself but emerges clearly with the help of a truth table. This, of course, is the purpose of such tables. The reader may care to construct a truth table for the compound statement $(p \wedge q) \vee (\sim p \wedge \sim q)$. The solution is shown in Table 4.

TABLE 4

p	q	$(p \wedge q) \vee (\sim p \wedge \sim q)$
T	T	T
T	F	F
F	T	F
F	F	T

If p and q relate to switches, however, and we consider when p and q have values 0 (open) and 1 (closed), then the values of $p \vee q$ and $p \wedge q$, etc., will be either 0 (no current) or 1 (current flows). We then have closure tables for the switching circuits (i) and (ii) (Table 5).

TABLE 5

p	q	$p \vee q$	$p \wedge q$	$\sim p \wedge \sim q$	$(p \wedge q) \vee (\sim p \wedge \sim q)$
1	1	1	1	0	1
1	0	1	0	0	0
0	1	1	0	0	0
0	0	0	0	1	1

Assertions are not always such bold statements as the ones discussed above. We may wish to couch an assertion in conditional terms: "If England wins the Test match I shall be happy." Reduced to symbolic form this is a statement of the type "if p then q" or $p \rightarrow q$. If p and q are both true then $p \rightarrow q$ is true, while if p is true and q is false $p \rightarrow q$ is obviously false. It is not quite clear, however, what happens if p is false. The only thing to do

APPLICATIONS OF SET THEORY   31

is to make the *definition that whenever p is false then q is true*. In other words, if England does not win the Test match it is true to say that I may or may not still be happy. We now have the truth table (Table 6).

TABLE 6

p	q	$p \to q$
T	T	T
T	F	F
F	T	T
F	F	T

Our final connective was $p \leftrightarrow q$ or "if p then q and if q then p". In this case the truth table is obvious; the statement can only be true if p and q are both true or both false. Thus $p \to q$ is true whenever $p \leftrightarrow q$ is true. We say that $p \leftrightarrow q$ *implies $p \to q$*.

So far we have used the symbol $\to$ to form a new statement from two given statements. The most interesting situation arises when considering the relation between two statements, and the most interesting case here is when one statement implies the other.

If the nature of two statements p and q is such that the only possible cases which can arise make $p \to q$ logically true (i.e. true in all cases that arise), then we say that *p implies q*. The statements p and q are said to be *equivalent* if $p \leftrightarrow q$ is logically true in all cases that arise.

A relation involving implication can often be simplified by means of a truth table. Consider, for example, the implication $(p \wedge \sim q) \to \sim p$. The truth table is as shown in Table 7.

TABLE 7

p	q	$\sim q$	$\sim p$	$(p \wedge \sim q)$	$(p \wedge \sim q) \to \sim p$
T	T	F	F	F	T
T	F	T	F	T	F
F	T	F	T	F	T
F	F	T	T	F	T

If we compare this with the truth table for $p \to q$ we see that the tables are identical and hence $(p \wedge \sim q) \to \sim p$ is equivalent to $p \to q$.

If Jack is a Yorkshireman p,
and Jack is an Englishman q,
then p clearly implies q.

$(p \wedge \sim q) \to \sim p$ makes the same point in a more obscure fashion: "If Jack is a Yorkshireman and not an Englishman then he is not a Yorkshireman."

Given any two statements comprising the same parts (p and q, etc.) it is not always possible to see whether they are consistent. If, however, we reduce each to a symbolic statement and compare the truth tables, the uncertainty is soon resolved.

ALTERNATIVE NOTATION

The function or statement $(p \wedge q) \vee (\sim p \wedge \sim q)$, although a very simple one, looks rather complicated to unfamiliar eyes. In this notation a complex statement can look quite formidable. For this reason other writers denote:

the negation of a by a'
the assertion of either a or b (or both) by $a+b$
the assertion of a and b by $a.b$,

so that the function $(p \wedge q) \vee (\sim p \wedge \sim q)$ becomes $p.q + p'.q'$ or even $pq + p'q'$. Expressed in this form functions are more easy to manipulate and an algebraic argument or solution has the virtue of looking much simpler. On the other hand, the signs + and . are so familiar that there is a danger of using them, perhaps unwittingly or "automatically", as we should do in ordinary algebra. This, however, is by no means ordinary algebra; it is an *algebra of sets*, or an example of what is known as *Boolean algebra* after Boole (1815–64).

If P and Q are the truth sets of statements p and q, p *and* q, $p \wedge q$ or $p.q$ is true in the truth set $P \cap Q$. Also, either p *or* q (*or both*), that is $p \vee q$ or $p+q$ is true in the set $P \cup Q$. Corresponding to the union ∪ we have the connective ∨ and the operation +, and to the intersection ∩ we have connective ∧ and the operation .. The laws of statements are exactly analogous with the laws of sets.

From this point onwards we shall use mainly the alternative notation.

BOOLEAN ALGEBRA

All Boolean algebras are laws of sets. To produce an algebra of statements (and, as we shall see, of switching circuits) we rewrite the laws of sets 1–12 deduced in the previous chapter replacing ∪ by + ; ∩ by . or × ; ∅ by 0 and $\mathscr{E}$ by 1.

We then have

1. $A \times 1 = A$
2. $A + 0 = A$
3. $A(B+C) = AB + AC$

These look familiar but

4. $A + BC = (A+B)(A+C)$
5. $A(A+B) = A$
6. $A + AB = A$
7. $A \times 0 = 0$
8. $A + 1 = 1$
9. $A \times A = A$
10. $A + A = A$
11. $(A+B)' = A'.B'$
12. $(AB)' = A' + B'$

This algebra incorporates an extraordinary *principle of duality*; for if × and + are interchanged, and 0 and 1 also, in any of the above relations, we obtain a relation which is still true.

SIMPLE APPLICATIONS TO LOGIC

The following examples illustrate how a logical argument or set of instructions may be simplified using this algebra.

Example

In a certain village

All sheep branded g belong to Farmer Giles,
Sheep do not wear collars unless they are branded g,
Farmer Giles has no black sheep.

Denote statements by letters as follows:

These sheep belong to Farmer Giles f,
These sheep are branded g g,
These sheep wear collars c,
These sheep are black b.

Then the original statements become

$$g \to f$$
$$c \to g$$
$$f \to b'.$$

If $\qquad g \to f$ and $f \to b'$, then $g \to b'$.
But $\qquad c \to g$, $\therefore c \to b'$ or $b.c = 0$,

and this statement reads: No black sheep wear collars (in this village).

Alternative solution:

$$gf' = 0, \qquad (1)$$
$$cg' = 0, \qquad (2)$$
$$fb = 0. \qquad (3)$$

Multiply 1 by b, $\qquad bgf' = b \times 0 = 0.$ (4)
Multiply 3 by g, $\qquad gfb = g \times 0 = 0.$ (5)
Add 4 and 5 $\qquad bg(f+f') = 0,$
but $\qquad f+f' = 1,$
$\qquad \therefore bg = 0.$ (6)
Multiply 6 by c, $\qquad cbg = c \times 0 = 0.$ (7)
Multiply 2 by b, $\qquad bcg' = b \times 0 = 0.$ (8)
Add 7 and 8 $\qquad bc(g+g') = 0,$
but $\qquad g+g' = 1,$
$\qquad \therefore b.c = 0,$

i.e. no black sheep wear collars.

Finally, we could deduce the main conclusion by means of a Venn diagram. If S represents the set of all sheep in the village, F the set of sheep which are branded g and belong to Farmer Giles, C the set of sheep wearing collars, and B the set of

APPLICATIONS OF SET THEORY

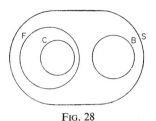

Fig. 28

black sheep, then the diagram will be as shown in Fig. 28, i.e. $B \cap C = \emptyset$ or, in the algebra of statements,

$$b.c = 0,$$

or, no black sheep wear collars.

Example

Confronted with the following set of regulations, a motorist, though anxious to conform with the law, was, nevertheless, understandably perplexed. Simplify them:

When you do not keep to the left, sound the horn,
If you keep to the left and sound the horn, do not stop,
If you are stationary, or you are on the right, do not sound the horn.

Denote "you keep to the left" by l,
"you should sound the horn" by h,
"you should stop" by s.

Then the statements reduce to

$$l' \to h$$
$$l \wedge h \to s'$$
$$s \vee l' \to h'$$

or, in the alternative form

$$l'h' = 0$$
$$lhs = 0$$
$$(s+l')h = 0.$$

Adding together we have

$$l'h' + lhs + (s+l')h = 0,$$
$$\therefore l'(h'+h) + hs(l+1) = 0.$$

Now $h'+h = 1$ and $l+1 = 1$,

$$\therefore l' + hs = 0 \qquad (1)$$

Remember now that

$$(A \cup B)' = A' \cap B' \text{ reduces to}$$
$$(a+b)' = a'.b'$$

so that the complement of $a+b$ is $a' \times b'$.
Take the complement of both sides of 1 and we have

$$l(hs)' = 1.$$

This reads "Always drive on the left and do not sound the horn when stationary", a rather simpler instruction to understand.

SIMPLE APPLICATIONS TO SWITCHING CIRCUITS

The Boolean algebra which we have just introduced and used can be further applied to the analysis of switching circuits. In the example quoted earlier in this chapter we had switches p and q (i) in series, (ii) in parallel. We saw that in the algebra of propositions we could write:

(i) $p \wedge q \to c$, and (ii) $p \vee q \to c$.

Expressing these propositions in Boolean algebraic form we have:

(i) $pq = 1$, (ii) $p+q = 1$

or

(i) if p is shut *and* q is shut current flows,
(ii) if either p *or* q is shut current flows.

If we examine the closure table given earlier (Table 5) showing values of $p \vee q$ ($p+q$) and $p \wedge q$ (pq) for all possible combinations of p and q on and off, we see that these functions hold in all cases.

In switching Boolean algebra we denote a switch by a single letter $a, b \ldots, x, y \ldots$. If two switches open and close

APPLICATIONS OF SET THEORY

simultaneously we denote them by the same letter. If one is open when the other is closed and vice versa we denote one by x and the other by x'. Switches x and y in *parallel* are denoted by $x+y$, and switches x,y in *series* are denoted by xy.

Using these rules we can build up a Boolean function for any circuit which involves sets of switches connected either in series or in parallel.

Examples

(i)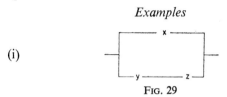

Fig. 29

Boolean function: $x+yz$ (Fig. 29).

(ii)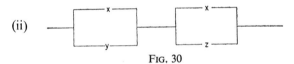

Fig. 30

$(x+y)(x+z)$ (Fig. 30).

(iii)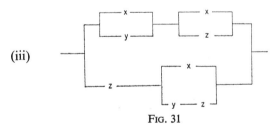

Fig. 31

$(x+y)(x+z)+z(x+yz)$ (Fig. 31).

(iv)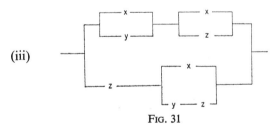

Fig. 32

$(x'+y')(x'+y)(x+y)$ (Fig. 32).

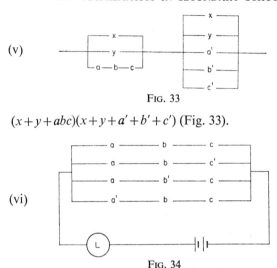

Fig. 33

$(x+y+abc)(x+y+a'+b'+c')$ (Fig. 33).

Fig. 34

$abc+abc'+ab'c+a'bc$ (Fig. 34).

Consider now the circuits given in examples (i) and (ii). The table of closure properties for these is shown in Table 8. As we

TABLE 8

x	y	z	Circuit (i)	Circuit (ii)
1	0	0	1	1
1	0	1	1	1
1	1	0	1	1
1	1	1	1	1
0	0	0	0	0
0	0	1	0	0
0	1	0	0	0
0	1	1	1	1

might expect the closure properties for the two circuits are identical, for their functions $x+yz$ and $(x+y)(x+z)$ are identical by rule 4 of the laws of Boolean algebra. We may, therefore, use any of the rules 1 to 12 to analyse or simplify a switching circuit.

Consider (iii).

APPLICATIONS OF SET THEORY

The Boolean function is

$$(x+y)(x+z)+z(x+yz)$$
or $\quad x+yz+z(x+yz)$ (by 4)
or $\quad (x+yz)(1+z)$ (by 3)
or $\quad x+yz$ (since $1+z = 1$) (by 8).

The circuit (iii) therefore has the same closure properties as circuit (i) (Fig. 29).

Consider now circuit (iv).

The Boolean function is

$$(x'+y')(x'+y)(x+y)$$
or $\quad (x'+y'y)(x+y)$ (by 4)
or $\quad x'(x+y)$ (since $y'y = 0$)
or $\quad x'y$ (since $x'x = 0$),

i.e. circuit (iv) reduces to the simpler circuit —x'—y———.

Circuit (v) has the Boolean function

$$(x+y+abc)(x+y+a'+b'+c').$$

Now by 12 $\qquad a'+b' = (ab)'$
and *in extenso* $\qquad a'+b'+c' = (abc)'.$

Thus the function reduces to

$$(x+y+abc)\{x+y+(abc)'\}$$
or $\quad \{x+y+(abc)(abc)'\}$ (by 4)
or $\quad x+y$ since $(abc).(abc)' = 0.$

Thus circuit (v) reduces to —[$\frac{x}{y}$]—.

Circuit (vi) deserves closer attention. If we write out the table of closure properties we have Table 9.

We notice that when $a+b+c \geq 2$, or when a majority of the switches are closed, then current flows and the lamp L lights up.

Consider now a sort of "Juke Box Jury" panel of three members. Each member votes a "hit" by pressing his switch or button (a, b or c), or records a "miss" by leaving the switch open. We require a circuit which enables a majority vote in favour of a "hit"

TABLE 9

a	b	c	Value of $abc + abc' + ab'c + a'bc$ or simply by inspection of the circuit
1	1	1	1
1	1	0	1
1	0	1	1
0	1	1	1
1	0	0	0
0	1	0	0
0	0	1	0
0	0	0	0

to be recorded either by a light or a buzzer. Obviously circuit (vi) is the solution to this problem; or rather it is one solution to the problem.

For $\quad abc + abc' + ab'c + a'bc$
simplifies to $\quad ab(c+c') + c(ab' + a'b)$ (by 3)
or $\quad ab + c(ab' + a'b)$ (since $c + c' = 1$).

Thus the circuit (vi) can be simplified as shown in Fig. 35.

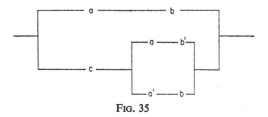

Fig. 35

It is the designing of a circuit from given data, rather than the simplification of circuits, which in practice is more useful. It is a simple matter to deduce the table of closure properties from a given function. Of greater value, however, is the ability to write down the closure properties required and from this deduce the function, and hence the circuit, which satisfies the given conditions.

APPLICATIONS OF SET THEORY

As an example we discuss the design of an ordinary household two-way switch by which a light may be switched either on or off from upstairs or downstairs.

Suppose we have a switch p upstairs and a switch q downstairs. These switches must control the circuit independently. Assume that in the initial state p and q are both down (i.e. both have value 1) and that the light is on (function has value 1). Successive single changes in p or q must create changes in the value of the function. The closure properties required are as shown in Table 10.

TABLE 10

	p	q	function
One change at a time	1	1	1
	0	1	0
	0	0	1
	1	0	0

Examining the table we see that the function has value 1 when p AND q are both down OR when p AND q are both up. Using symbols, the function has the same value for $(p \wedge q) \vee (\sim p \wedge \sim q)$ or as a Boolean function the circuit required may be expressed $pq + p'q'$. (Earlier in the chapter we deduced the closure properties of $(p \wedge q) \vee (\sim p \wedge \sim q)$ as an exercise and, of course, we obtained the table shown above.) The required

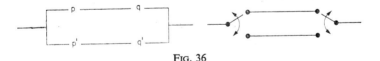

Fig. 36

circuit is shown in Fig. 36, i.e. we require a single pole two-way switch on each floor.

Consider now a long corridor illuminated by a strip filament. Design a circuit by which the light may be independently controlled from any one of three switches placed at the middle and both ends of the corridor.

Denote the switches by letters p, q and r. The closure table for all possible values of p, q and r, satisfying the requirements that for a change in any one switch there must be a change in the Boolean function of the circuit, is as shown in Table 11.

TABLE 11

p	q	r	circuit
0	0	0	0
1	0	0	1
0	1	0	1
0	0	1	1
1	1	0	0
1	0	1	0
0	1	1	0
1	1	1	1

Current flows when

p and q and r all have value 1
OR
p' and q' and r all have value 1
OR
p' and q and r' all have value 1
OR
p and q' and r' all have value 1.

Thus the Boolean function of the circuit required is

$(p \wedge q \wedge r) \vee (\sim p \wedge \sim q \wedge r) \vee (\sim p \wedge q \wedge \sim r) \vee (p \wedge \sim q \wedge \sim r)$
or $\qquad pqr + p'q'r + p'qr' + pq'r'$

or the circuit shown in Fig. 37.

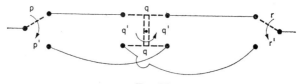

Fig. 37

For four independent switches the Boolean function is

$$pqrs + \sum p'q'rs + p'q'r's'$$

for five the Boolean function is

$$pqrst + \sum p'q'rst + \sum p'q'r's't$$

and so on.

Finally, we sometimes require circuits with built-in conditions. Reverting to the three-man committee of example (vi), let us suppose that for the light to go on and indicate a "hit",

(a) a majority vote is required,
(b) a hit is not recorded unless the chairman's vote is in favour. Assuming that the chairman controls switch a, the closure table is as shown in Table 12.

TABLE 12

a	b	c	L
1	1	1	1
1	1	0	1
1	0	1	1
1	0	0	0
0	1	1	0
0	1	0	0
0	0	1	0
0	0	0	0

The light goes on for

a and b and c

OR

a and b and c'

OR

a and b' and c.

The Boolean function required is therefore

$$(a \wedge b \wedge c) \vee (a \wedge b \wedge \sim c) \vee (a \wedge \sim b \wedge c)$$
or
$$abc + abc' + ab'c$$

or Fig. 38.

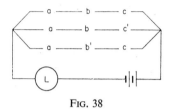

Fig. 38

This result is capable of considerable simplification, for

$$\begin{aligned}
& abc + abc' + ab'c \\
&= ab(c+c') + ab'c \\
&= ab + ab'c \quad \text{(since } c+c' = 1\text{)} \\
&= a(b + b'c) \\
&= a(b+b')(b+c) \quad \text{(by law 4)} \\
&= a(b+c) \quad \text{(since } b+b' = 1\text{)}.
\end{aligned}$$

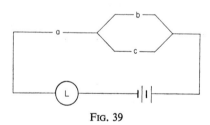

Fig. 39

Hence the simplest circuit which will satisfy the requirements is shown in Fig. 39.

Useful Reference Books

ALLEN, R. G. D., *Basic Mathematics*, Macmillan.
BAILEY, C. A. R., *Sets and Logic I and II*, E. Arnold.
BOWRAN, A. P., *A Boolean Algebra*, Macmillan.

FLETCHER, T. J., *Some Lessons in Mathematics*, C.U.P.
GOODSTEIN, R. L., *Boolean Algebra*, Pergamon.
GOODSTEIN, R. L., *Fundamental Concepts of Mathematics*, Pergamon.
KEMENY, J. G., MIRKIL, H., SNELL, J. L. and THOMPSON, G. L., *Finite Mathematical Structures*, Prentice-Hall.
KURATOWSKI, K., *Introduction to Set Theory and Topology*, Pergamon.
MOAKES, A. J., *The Core of Mathematics*, Macmillan.
RICHARDSON, M., *Fundamentals of Mathematics*, Macmillan, New York.
WHITESITT, E., *Boolean Algebra and its Applications*, Addison-Wesley.

3
GROUPS

IN THE last two chapters we showed how it was possible, starting with the simple idea of a set, to build a mathematical structure and to use the algebra of this structure in the solution of certain types of problem. As far as the algebra of sets is concerned we are interested in the relationships between sets, their intersections, unions and complements; we are not concerned with relationships between the elements of those sets except in so far as these elements are well-defined objects, operations or numbers of a similar type.

In this chapter we consider an extremely important subset of the set of all sets in which certain relationships between the elements do exist. Nevertheless, this class of sets, which we shall call *groups*, is a very wide one and the group concept, as we shall see, is an important unifying principle in many branches of mathematics.

Before defining a group we shall discuss symmetry and explain what is meant by a binary operation.

BINARY OPERATIONS

A *binary operation* is one by which a member of a set is obtained by the *combination* of two other members of that set. This combination may take several forms. For example we may add together or multiply together two members of a set to produce a third (but not necessarily different) member of the set. In a set of rotations a binary operation would consist in applying one rotation, following it by a second to produce a third rotation. In sets of transformations or permutations, the binary operation is simply the successive application of two transformations or permutations to produce a third transformation or permutation which is also a member of the set.

Examples of binary operations

$2+3 = 5$; $2 \times 3 = 6$; rotation 30° followed by rotation 40° ≡ rotation 70° (assumed coplanar and in the same sense); permutation $\binom{a}{b}$ operating on permutation $\binom{b}{c}$ produces permutation $\binom{a}{c}$; translation $A \to B$ followed by translation $B \to C$ equivalent to translation $A \to C$.

Under *vector* multiplication, the product of two vectors produces a third vector, i.e. $\mathbf{a} \times \mathbf{b} = \mathbf{c}$; and in a very broad sense even, $H_2O + SO_3 = H_2SO_4$, where two molecules combine to form a third member of the set of all molecules.

Examples which are not binary operations

The *scalar* product of two vectors is a pure number: $\mathbf{a}.\mathbf{b} = ab \cos.\theta$ where θ is the angle between $\mathbf{a}$ and $\mathbf{b}$.

$Na + I = NaI$, where atom combines with atom to produce a molecule.

And, if we consider only the set of negative integers,

$$-2 \times -3 = +6,$$

for $+6$ is not a member of that set.

Usually we consider a set from the point of view of one operation; we talk of a "group under addition" or a "group under multiplication" or a "group of rotations", etc. Some sets, however, are groups under two operations; these we describe as systems of *double composition*.

SYMMETRY

Most people have some notion of symmetry. It is an idea which is linked with order and beauty. We speak of symmetrical features. Gardens and rooms at one time were frequently laid out or set out in a symmetrical form. The average person will react to the word pentagon by visualizing a regular pentagon; small boys asked to sketch a triangle will, more often than not, sketch an isosceles or even an equilateral triangle. It is an idea which forms in the minds of children at quite an early age. We have all met small children who, in some respect or other, are very neat and orderly; order and symmetry seem to play some essential part in their feelings of security. And yet, until recent years, this concept of symmetry is one which has not been encouraged or developed in school mathematics.

Some people do have the hazy idea that symmetry is in some way linked with geometry, but then how does one "measure" symmetry? At the more sophisticated stage of sixth-form mathematics we expect our pupils to recognize symmetry in algebraic expressions, in problems on frameworks, etc., and to use this symmetry to advantage in producing elegant solutions. The better mathematicians of course do, but one often feels that they do in spite of their previous training. One remembers having to learn by heart a proof by congruence of the fact that the base angles of an isosceles triangle are equal. The fact seemed perfectly obvious, almost axiomatic, at that age, but arguments based on symmetry or folding or rotation about the altitude, were regarded as heresy and henceforth discouraged. Recently, geometry-teaching has become far more enlightened, but one feels that there are still steps to be taken in the proper teaching of symmetry.

In the first place there are many different kinds of symmetry, and, indeed, much group theory is concerned with the groups of symmetries of geometrical figures, algebraic and differential equations, or any situation in which patterns of symmetry exist. It is often possible to analyse such groups into simpler subgroups, and in this way make conclusions or solve problems. At an advanced level group theory has been applied in quantum physics to the problem of determining molecular structure in chemistry; it is being used now in the field of mathematical biology to investigate patterns of symmetry occurring in genetic mutations. Applications at this level involve extremely difficult mathematics, but the foundations can be laid very early in the secondary school course.

Consider for a moment what we mean when we expect a sixth-form boy to recognize symmetry in a mathematical situation. Here are a few common examples:

(i) Find the coordinates of the point G which lies two-thirds of the way from A along the median AD in the triangle $A(x_1 y_1)$, $B(x_2 y_2)$, $C(x_3 y_3)$. Prove that the other medians also pass through this point.

The point G is

$$\left(\frac{x_1+x_2+x_3}{3}, \frac{y_1+y_2+y_3}{3}\right)$$

The concurrence of the medians follows immediately from the symmetry of the coordinates of G. For replacing x_1 by x_2, x_2 by x_3, etc., we obtain the corresponding point on another median and G is unaltered. There is a cyclic symmetry of the form

(ii) Factorize

$$(b-c)^3+(c-a)^3+(a-b)^3.$$

The expression vanishes for $b = c$, therefore $(b-c)$ is a factor. By symmetry $(c-a)$ and $(a-b)$ are also factors, etc. Here we have a cyclic symmetry of the same form

(iii) The same type of symmetry yields an elegant solution when factorizing

$$\begin{vmatrix} a & b & c \\ a^2 & b^2 & c^2 \\ a^3 & b^3 & c^3 \end{vmatrix}$$

or when determining $\sum \alpha^2$, where α, β, γ are roots of $x^3+p_1 x^2+p_2 x+p_3 = 0$.

(iv) In problems where complex roots are involved, replacement of i by $-$i may reveal the presence of conjugate pairs. In other words an element of symmetry about the real axis in the Argand plane may be involved. This is a different kind of symmetry from that in (i)–(iii).

(v) The "quick" way of evaluating

$$\int_0^{\frac{\pi}{2}} \cos^2 x \, dx$$

by writing

$$\int_0^{\frac{\pi}{2}} \cos^2 x \, dx = \int_0^{\frac{\pi}{2}} \sin^2 x \, dx$$
$$= \frac{1}{2} \int_0^{\frac{\pi}{2}} (\cos^2 x + \sin^2 x) \, dx = \frac{\pi}{4}$$

is based on the symmetry of the function $\cos x$ about the line $x = 0$. It is a similar kind of symmetry to that in (iv).

(vi) The proof that the diagonals of a regular hexagon are concurrent is based on the group of permutations of the vertices of the figure. The regular hexagon may be rotated through an angle $\frac{1}{3}\pi$ in its plane, about the midpoint of any diagonal, without altering its position or appearance, and this can be done six times. A cyclic group of rotations of order six is involved in this type of symmetry.

(vii) A symmetrical framework of light rods, symmetrically loaded, must give rise to a symmetrical force diagram, and for $2n$ vertices the force diagram is complete by the time the first n vertices have been dealt with.

THE IDEA OF A GROUP

In practice, of course, pupils do not always recognize the existence of these symmetries.

A deliberate mathematical approach to symmetry can be made in the first year of the secondary school course. One way of doing this is as follows.

Consider the capital letters of the alphabet. Some of them, such as F and G, are not symmetrical. Some, like A and M, are symmetrical and so are B and K. The symmetry of A, however, is about a vertical axis, while that of B is about a horizontal axis. After a rotation through 180°, or a "*half-turn*" *about a horizontal axis*, the letter B is unaltered; call such a rotation p. For A to remain unaltered in position and appearance, we require a *half-turn about the vertical axis*; call this q.

There is a third type of symmetry associated with the letter N. The operation p on N produces И; the operation q on N produces the same thing — И. If we follow an operation p on N by an operation q we do reproduce N. But there is a single operation which will produce N; a rotation through 180° or a *half-turn in the plane of the paper*, i.e. about an axis perpendicular to the paper. By such an operation N becomes N, or, as it appears, upside down. We call this type of rotation r. It follows that operation q followed by operation p has the same effect as operation r. Let us write $p.q = r$. Note further that in this particular case $q.p = r$ also. We can find other relations between p, q, r such as $p.r = q$, but we shall not pursue this at the present stage.

If we look at the alphabet again we find that there are other even more symmetrical letters. H, I, O and X are all unchanged by any of the operations p, q and r. It is a good simple exercise to examine all the letters of the alphabet and to decide which sort of symmetry applies in each case. In fact all the letters, except F, G, J, L, P, Q and R, remain unchanged after one or other or all three of the operations p, q, r. In order to bring these letters into the scheme we introduce a rather trivial operation I called the identity operation, which stands for "leave the letter as it is". Thus, in the case of N, r or $p.q$ has the same effect as I. Note also that the result of the operation p can be undone by a further application of the operation p; $p.p$ or $p^2 = I$. When the effect of an operation x can be reversed by a further operation y, we say that operation y is the *inverse* of operation x. As far as N is concerned p is its own inverse and q is its own inverse, but since $p.q = I$, q is also an inverse of p. This is to say that as far as N is concerned, the set of operations I, p, q, r does not contain a *unique identity element* (r has the same effect as I), and each element does not possess a *unique inverse element* (p and q are both inverses of p itself). We shall see, however, that there are sets of operations applied to certain figures where uniqueness does exist and it is such cases which are really interesting.

Consider now the problem of fitting a rectangular pane of glass into a rectangular window frame of the same dimensions. In how many ways can we place the glass in position? This can be done practically by children using rectangles of clear acetate paper marked with chinagraph pencils. A highly suitable system of lettering is H, I, O, X. [These letters look the same after any rotation through a half-turn, or viewed from either side of the sheet.] The possible ways, together with the linking operations are shown in Fig. 40.

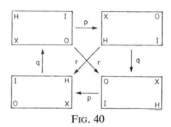

Fig. 40

Thus, for example,
$$p^2 = q^2 = r^2 = I, \quad p.q = q.p = r,$$
$$p.r = r.p = q, \quad q.r = r.q = p,$$

and, of course,

$$I^2 = I, \quad Ip = p, \quad Iq = q \quad \text{and} \quad Ir = r.$$

We can set up these results in the form of a "multiplication" table (Table 13).

TABLE 13

	I	p	q	r
I	I	p	q	r
p	p	I	r	q
q	q	r	I	p
r	r	q	p	I

Notice that in the case of the rectangle HIOX the set of elements $\{I, p, q, r\}$ contains:

(i) A *closure* property or *binary operation*, i.e. the "product" of any two elements is a member of the set.

(ii) *A unique identity element I.*
(iii) *A unique inverse element* for each element of the set (*I* appears once only in any row or column). Finally—
(iv) An *associative law* holds. That is to say any expression of the type $a \times (b \times c)$ is equal to $(a \times b) \times c$,
e.g. $(q \times r) = p$, therefore $p \times (q \times r) = p \times p = I$,
also $(p \times q) = r$, therefore $(p \times q) \times r = r \times r = I$,
and so on for all possible trios of elements.

As we shall see, (i), (ii), (iii), and (iv) *are the necessary conditions for the set to be a group.* We call this set of operations $\{I, p, q, r\}$ the *group of the rectangle.*

Generally speaking the more symmetrical the figure, the larger the group. N is not a very symmetrical letter. It has a very small group of symmetries $\{I, r\}$ with the multiplication table

$$\begin{array}{c|cc} & I & r \\ \hline I & I & r \\ r & r & I \end{array}$$

The letter G is not regenerated at all except by the operation I; it has the trivial group

$$\begin{array}{c|c} & I \\ \hline I & I \end{array}$$

On the other hand, had our sheet of glass been square instead of rectangular, it would have been symmetrical under quarter-turns as well as half-turns. Starting with the square ABCD, this square can be placed into the window in eight different ways:

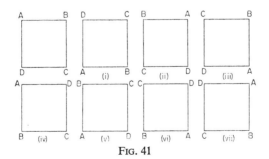

FIG. 41

These different ways are a result of the following operations:

(i) A half-turn about a horizontal axis (operation p).
(ii) A half-turn about a vertical axis (operation q).
(iii) A half-turn about the diagonal DB (operation r).
   [Note that here r is defined differently, the previous definition gives (vi) not (iii).]
(iv) A half-turn about the diagonal AC (operation s).
(v) A counter-clockwise rotation through one right-angle (operation i).
(vi) A counter-clockwise rotation through two right-angles (operation -1).
(vii) A counter-clockwise rotation through three right-angles (operation $-i$).

The original figure may be regarded as a counter-clockwise rotation through zero angle, the identity operation 1. The relationships between these operations may again be expressed in the form of a "multiplication" table (Table 14).

TABLE 14

×	1	i	−1	−i	p	q	r	s
1	1	i	−1	−i	p	q	r	s
i	i	−1	−i	1	r	s	q	p
−1	−1	−i	1	i	q	p	s	r
−i	−i	1	i	−1	s	r	p	q
p	p	s	q	r	1	−1	−i	i
q	q	r	p	s	−1	1	i	−i
r	r	p	s	q	i	−i	1	−1
s	s	q	r	p	−i	i	−1	1

First operation is along the top; Second operation is along the side.

If we examine Table 14 we find that the closure and associative properties hold. There is a unique identity element 1, and to each element there corresponds one, and only one, inverse element. The set of operations $\{1, i, -1, -i, p, q, r, s\}$ is called the *group of the square*. It is a large group and this is to be expected from the very symmetrical nature of the square.

Closer examination reveals further interesting features of this group. Whereas the multiplication table of the group of the rectangle is symmetrical about its diagonal, i.e. it is a *commutative*

GROUPS 55

or *Abelian group* with $p \times q = q \times p$, etc., we find that in this case the group of the square is not commutative. For example, operation p followed by operation i, written ip, is operation r; operation i followed by operation p, written pi, is operation s.

Further, we notice that this large group of the square contains smaller groups within itself. The multiplication table contains the following subsidiary tables.

TABLE 15

$\times$	1	i	-1	$-$i
1	1	i	-1	$-$i
i	i	-1	$-$i	1
-1	-1	$-$i	1	i
$-$i	$-$i	1	i	-1

(a) A group of *order 4*—or the group of quarter-turns of the square (Table 15). The reader conversant with complex numbers will see at once why we have denoted the quarter-turn operation

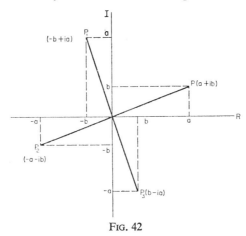

FIG. 42

in this way. If the complex number $a+ib$ ($i = \sqrt{-1}$; a, b real) is represented by the vector $\overrightarrow{OP}$ in the Argand plane, $i.\overrightarrow{OP}$ or $\overrightarrow{OP_1}$ represents $i(a+ib)$ or $ia-b$ and $i.\overrightarrow{OP}$ *is the vector $\overrightarrow{OP}$ rotated counter-clockwise through one right-angle*. This is seen from Fig. 42.

Further multiplication of $\overrightarrow{OP_1}$ by i produces $\overrightarrow{OP_2}$, i.e. another quarter-turn. $\overrightarrow{OP_2} = i.\overrightarrow{OP_1} = i \times i.\overrightarrow{OP} = -1.\overrightarrow{OP}$ and the second operation, -1, clearly produces two quarter-turns of $\overrightarrow{OP}$. Finally, $-i$ sends $\overrightarrow{OP}$ to $\overrightarrow{OP_3}$ through three quarter-turns.

In the present context there is no need to use 1, i, -1, and $-i$ for the quarter-turn operations, any other four symbols would have served equally well. On the other hand, there is no point in obscuring the connexion between quarter-turns and the function of the operator i in the Argand plane. Links such as this illuminate and help to unify the various concepts. (In our next chapter we shall see that a quarter-turn in the Cartesian plane is effected by the operation of a second-order square matrix.)

(b) Another smaller group is revealed if we pick out the elements 1, p, q, -1 and the products corresponding to them (Table 16).

TABLE 16

×	1	p	q	-1
1	1	p	q	-1
p	p	1	-1	q
q	q	-1	1	p
-1	-1	q	p	1

If we replace -1 by r as defined in the group of the rectangle we have Table 17, which is the group of the rectangle, i.e. the

TABLE 17

×	I	p	q	r
I	I	p	q	r
p	p	I	r	q
q	q	r	I	p
r	r	q	p	I

group of the square contains the group of the rectangle, a result which we might well have expected.

GROUPS

(c) The groups $\{1, p\}$, $\{1, q\}$, $\{1, -1\}$

×	I	p
I	I	p
p	p	I

which are the groups of rotations of the letters B, A, N respectively.

Cases (a), (b) and (c) are called SUBGROUPS of the group of the square. Notice that not all subsets of $\{1, i, -1, -i, p, q, r, s\}$ are subgroups. In other words a group *may* possess subgroups. A group very frequently possesses subsets which are not groups. On the other hand, we come across cases of sets which are not groups, possessing proper subsets which are. Subgroups will be discussed a little more fully at a later stage in this chapter.

We now come to the essential point in the idea of a group. We chose to develop the group of the square by discussing the fitting of a square piece of glass into a frame. Had we discussed the ways in which an X-shaped girder could be placed in position we should have obtained precisely the same group of symmetries. In other words, the square, the X-shaped girder, the rectangle and the letters we discussed are *essentially* of no interest to us; our exclusive concern lies with the groups of symmetries, the patterns associated with these objects. Group theory as it were, takes a set of objects, elements or operations and extracts only the intrinsic symmetries for further analysis.

DEFINITION OF A GROUP

A set $S = \{a, b, c, \ldots\}$, consisting of either a finite or an infinite number of elements of any specified kind in which a law of combination (a binary operation) ∗ is specified, is a group if the following postulates are satisfied:

(i) a closure property exists, i.e. for all elements of S such as $a, b, a * b \in S$ (actually this is implied in the definition of a binary operation; we include it here merely to remind the reader of that definition),

(ii) an associative property exists, i.e. $a * (b * c) = (a * b) * c$ where a, b, c are elements of S not necessarily all different,

(iii) there exists a unique identity element, i.e. we can find $e \in S$ such that $e * a = a$ for all elements of S such as a,

(iv) there exists an inverse element, i.e. for each element of S such as a there exists $a^{-1} \in S$ such that $a^{-1} * a = e$.

A group does not necessarily possess a commutative property. If, however, for all elements of S such as a and b, $a * b = b * a$, the group is called commutative or Abelian.

GROUPS UNDER ADDITION

If the rule of combination is ordinary arithmetical addition the definition requires that

(i) $a+b$ is a member of the group,
(ii) $a+(b+c) = (a+b)+c$,
(iii) $o+a = a$ (the identity element is zero),
(iv) $(-a)+a = 0$ (the inverse of a is $-a$).

Examples

1. The set J of integers is a commutative group under addition.

$$2+3 = 3+2 = 5 \quad \text{a member of the group}$$
$$2+(3+4) = (2+3)+4$$
$$0+2 = 2$$
$$-2+2 = 0$$

(Note that J^+, the set of positive integers, is not a group under addition since $-2 \notin J^+$.)

2. The set of integers mod n, e.g. $\{0, 1, 2, 3, 4\}$ mod 5.

Note.
Counting in the scale of 10 we have
   1, 2, 3, 4, 5, 6, 7, 8, 9, 10, 11, 12, ...
Counting in the scale of 5 we have
     1, 2, 3, 4, 10, 11, 12, 13, 14, 20, 21, 22, ...
The "integers modulo 5" are
     1, 2, 3, 4, 0, 1, 2, 3, 4, 0, 1, 2, ...

GROUPS 59

[This is rather like taking the angle 370° as 10°, or the angle 730° as 10°.]

Practice in forming these tables in the classroom in various scales of notation is a useful way of re-examining the basic principles of arithmetic in a new light. It can also be tied up with the idea of groups.

The addition table for the integers mod 5 is as Table 18.

TABLE 18

+	0	1	2	3	4
0	0	1	2	3	4
1	1	2	3	4	0
2	2	3	4	0	1
3	3	4	0	1	2
4	4	0	1	2	3

The reader may like to verify that the following are also infinite groups under addition: the $\pm$ rational numbers, the $\pm$ real numbers, the complex numbers.

GROUPS UNDER MULTIPLICATION

If the rule of combination is ordinary arithmetical multiplication the definition requires that

(i) $a \times b$ is a member of the group,
(ii) $a \times (b \times c) = (a \times b) \times c$,
(iii) $1 \times a = a$ (the identity element is unity),
(iv) $\frac{1}{a} \times a = 1$ $\left(\text{the inverse of } a \text{ is } \frac{1}{a} \text{ or } a^{-1}\right)$.

Examples

1. $\left\{\ldots \frac{1}{2^3}, \frac{1}{2^2}, \frac{1}{2}, 1, 2, 2^2, 2^3, \ldots\right\}$

is a group under multiplication but not under addition, for

$$2^n \times 2^m = 2^{m+n}$$

which is an element of the group for all integral m, n. (Closure)

$$2^p \times (2^q \times 2^r) = (2^p \times 2^q) \times 2^r = 2^{p+q+r}$$

all integral p, q, r. (Associative property)

$$2^r \times 1 = 2^r$$

all integral r, therefore 1 or 2^0 is the identity element.

To each 2^s there corresponds an inverse 2^{-s} for all integral s.

2. The integers mod n (excluding 0) form a group under multiplication if n is prime, e.g. if $n = 5$ the multiplication table is as Table 19.

TABLE 19

×	1	2	3	4
1	1	2	3	4
2	2	4	1	3
3	3	1	4	2
4	4	3	2	1

Examination of this reveals that all the group properties are satisfied. The identity element is 1, the inverse of 2 is 3 and so on. If $n = 4$ however, the multiplication table is as Table 20.

TABLE 20

×	1	2	3
1	1	2	3
2	2	0	2
3	3	2	1

Note that $2 \times 2 = 0$, which is not a member of the group. Further 2 has no inverse.

∴ the integers mod 4 (excluding 0) do not form a group.

The reader may care to verify that the following infinite sets are also groups under *multiplication*: the ± rational numbers (excluding 0), the positive rational numbers, the ± real numbers (excluding 0), the complex numbers (excluding 0).

Note that the positive rational numbers form a subgroup of the ± rational numbers (excluding 0) and that the ± rational numbers (excluding 0) in turn form a subgroup of the $\pm$ real numbers (excluding 0). However, the set of integers $\backslash^0$, although a subset of the real numbers $\backslash^0$ ($\backslash^0 \equiv$ excluding 0), is not a subgroup of that group. On the other hand, the set of integers $\backslash^0$, while not a group, contains the subset $(1, -1)$ which is a group under multiplication.

GROUPS UNDER OTHER LAWS OF COMBINATION

We led up to the idea of a group by considering successive operations upon letters of the alphabet, the rectangle and the square. The reader should check that the group of the rectangle and the group of the square are in fact groups under the formal definition. We also used the term "multiplication" and the notations $\times$ and . to indicate that one operation was being followed by another. In all the following examples the term multiplication is used in this broad sense for want of a better.

In addition to the groups of rotations of the rectangle and the square, the following are also groups of operations:

1. The group of rotations of the equilateral triangle about its centroid (Fig. 43).

FIG. 43

Denote an anti-clockwise rotation of $2\pi/3$ radians about 0 by ω, of $4\pi/3$ by ω^2, of 0 radians by 1.

Thus (Fig. 44)

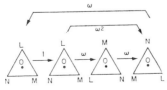

FIG. 44

and we find that $\{1, \omega, \omega^2\}$ forms a group with the multiplication table (Table 21).

TABLE 21

×	1	ω	ω^2
1	1	ω	ω^2
ω	ω	ω^2	1
ω^2	ω^2	1	ω

Incidentally, this is an example of a *cyclic group*.

2. $\{f_1, f_2, f_3, f_4\}$, where $f_1(z) = z$, $f_2(z) = -z$, $f_3(z) = 1/z$, $f_4(z) = -1/z$ and by multiplication we mean successive substitution. For example, f_3 means replace z by $1/z$, f_2 means replace z by $-z$, i.e. replace $1/z$ by $-1/z$.

$$\therefore f_2 f_3 \text{ means replace } z \text{ by } -1/z, \quad \text{which is } f_4.$$

Similarly, $f_3 f_4 = f_2$, etc. The identity element is, of course, f_1. The multiplication table is as Table 22.

TABLE 22

×	f_1	f_2	f_3	f_4
f_1	f_1	f_2	f_3	f_4
f_2	f_2	f_1	f_4	f_3
f_3	f_3	f_4	f_1	f_2
f_4	f_4	f_3	f_2	f_1

The following are also examples of groups under successive operations:

3. The group of rigid motions in three dimensions.

4. The group of rigid motions in three dimensions keeping one point fixed.

5. The group of translations in three dimensions.

ISOMORPHIC GROUPS

The process of abstracting the characteristic group of symmetries possessed by some sets of elements sometimes reveals a remarkable similarity of symmetrical structure between two sets of quite dissimilar objects or operations. As an example, let us recall the groups $\{1, i, -1, -i\}$ and $\{1, 2, 4, 3\}$ integers \\$^0$ mod 5. Their multiplication tables were as Tables 23 and 24.

TABLE 23

×	1	i	−1	−i
1	1	i	−1	−i
i	i	−1	−i	1
−1	−1	−i	1	i
−i	−i	1	i	−1

TABLE 24

×	1	2	4	3
1	1	2	4	3
2	2	4	3	1
4	4	3	1	2
3	3	1	2	4

If we make the substitutions $1 = 1$, $i = 2$, $-1 = 4$, $-i = 3$ in Table 23 we obtain Table 24. In other words these groups differ only in respect of notation; their *structures* are identical. Such groups are called *isomorphic groups*.

Another example is provided by the group of rotations of the equilateral triangle $(1, \omega, \omega^2)$ and the integers mod 3 under addition (Tables 25 and 26).

TABLE 25

	1	ω	ω^2
1	1	ω	ω^2
ω	ω	ω^2	1
ω^2	ω^2	1	ω

TABLE 26

+	0	1	2
0	0	1	2
1	1	2	0
2	2	0	1

The reader may like to check that the group of substitutions $\{f_1, f_2, f_3, f_4\}$, discussed as an example in the previous section, is isomorphic with the group of primes $\{1, 3, 5, 7\}$ mod 8 under multiplication and reduction as before.

Another way of putting the idea of isomorphism is to say that if G and G' are isomorphic groups, there is a *1:1 correspondence* or *mapping* of every element of G on to every element of G'. The mapping of G on to G' can be uniquely reversed; no two elements have the same image. If $G = \{a, b, \ldots\}$ and $G' = \{a', b', \ldots\}$ then the 1:1 correspondence $a \leftrightarrow a'$, $b \leftrightarrow b'$, ... exists such that $ab = c$ (for all a, b, c in G) implies $a'b' = c'$ for all a', b', c', in G'.

PERMUTATION GROUPS

We have discussed the group of rotations $\{1, \omega, \omega^2\}$ of an equilateral triangle ABC in its plane (Fig. 45). Had we considered the further motions p, a half-turn about the perpendicular from

Fig. 45

A to BC q, a half-turn about the perpendicular from B to AC, and r, a half-turn about the perpendicular from C to AB, we should have obtained the complete group of symmetries of the equilateral triangle in the same way as we obtained the complete group of the square. The reader may care to verify that this group has the multiplication table shown as Table 27.

TABLE 27

	×	1	ω	ω^2	p	q	r	Second rotation
	1	1	ω	ω^2	p	q	r	
	ω	ω	ω^2	1	q	r	p	
First rotation	ω^2	ω^2	1	ω	r	p	q	
	p	p	r	q	1	ω^2	ω	
	q	q	p	r	ω	1	ω^2	
	r	r	q	p	ω^2	ω	1	

We now discuss this problem from a different point of view. The problem of determining the number of ways in which an equilateral triangle may be placed into a frame of the same shape

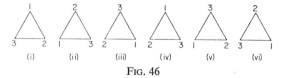

Fig. 46

is the same problem as that of determining the number of ways in which the letters of the vertices may be permutated amongst themselves. Denote the vertices A, B, C by the integers 1, 2, 3. The possible permutations are shown in Fig. 46. We denote the

GROUPS

change from (i) to (ii) by the notation $\begin{pmatrix} 1 & 2 & 3 \\ 2 & 3 & 1 \end{pmatrix}$ which means "change the 1 to 2, the 2 to 3 and the 3 to 1". In (iii), for example, 1 has become 3, 2 has become 1 and 3 has become 2, so the permutation by which (iii) is obtained from (i) is $\begin{pmatrix} 1 & 2 & 3 \\ 3 & 1 & 2 \end{pmatrix}$. If (i) remains unchanged after a permutation then that permutation can only be $\begin{pmatrix} 1 & 2 & 3 \\ 1 & 2 & 3 \end{pmatrix}$, the *identity permutation*. It will easily be seen that the cases (i) to (vi) respectively are produced from (i) by the following permutations:

$$\begin{pmatrix} 1 & 2 & 3 \\ 1 & 2 & 3 \end{pmatrix} \begin{pmatrix} 1 & 2 & 3 \\ 2 & 3 & 1 \end{pmatrix} \begin{pmatrix} 1 & 2 & 3 \\ 3 & 1 & 2 \end{pmatrix} \begin{pmatrix} 1 & 2 & 3 \\ 1 & 3 & 2 \end{pmatrix} \begin{pmatrix} 1 & 2 & 3 \\ 3 & 2 & 1 \end{pmatrix} \begin{pmatrix} 1 & 2 & 3 \\ 2 & 1 & 3 \end{pmatrix}$$

or by

$$p_0 \quad\quad p_1 \quad\quad p_2 \quad\quad p_3 \quad\quad p_4 \quad\quad p_5$$

By the "multiplication" of p_2 by p_3 we simply mean the permutation p_2 followed by the permutation p_3. Thus

$$\begin{pmatrix} 1 & 2 & 3 \\ 3 & 1 & 2 \end{pmatrix} \times \begin{pmatrix} 1 & 2 & 3 \\ 1 & 3 & 2 \end{pmatrix}$$

means change the 1 to 3, then this 3 to 2, i.e. change 1 to 2, change the 2 to 1, then this 1 to 1, i.e. change 2 to 1, change the 3 to 2, then this 2 to 3, i.e. change 3 to 3. So that

$$\begin{pmatrix} 1 & 2 & 3 \\ 3 & 1 & 2 \end{pmatrix} \times \begin{pmatrix} 1 & 2 & 3 \\ 1 & 3 & 2 \end{pmatrix} = \begin{pmatrix} 1 & 2 & 3 \\ 2 & 1 & 3 \end{pmatrix}$$

Or, by rearranging the second permutation and "cancelling" (we are not cancelling at all, but this is a convenient, quick way of "multiplying" permutations), we have

$$\begin{pmatrix} 1 & 2 & 3 \\ \cancel{3 \; 1 \; 2} \end{pmatrix} \times \begin{pmatrix} \cancel{3 \; 1 \; 2} \\ 2 & 1 & 3 \end{pmatrix} = \begin{pmatrix} 1 & 2 & 3 \\ 2 & 1 & 3 \end{pmatrix} \quad \text{as before.}$$

Thus $$p_2 \times p_3 = p_5.$$

This operation is not commutative for

$$p_3 \times p_2 = \begin{pmatrix} 1 & 2 & 3 \\ 1 & 3 & 2 \end{pmatrix} \times \begin{pmatrix} 1 & 3 & 2 \\ 3 & 2 & 1 \end{pmatrix} = \begin{pmatrix} 1 & 2 & 3 \\ 3 & 2 & 1 \end{pmatrix} = p_4.$$

However, the closure and associative properties apply; there is an identity element p_0 and each permutation possesses an inverse (e.g. the inverse of p_2 is p_1). Thus these permutations form a group and its multiplication table is given as Table 28.

TABLE 28

	×	p_0	p_1	p_2	p_3	p_4	p_5	Second permutation
	p_0	p_0	p_1	p_2	p_3	p_4	p_5	
	p_1	p_1	p_2	p_0	p_4	p_5	p_3	
First permutation	p_2	p_2	p_0	p_1	p_5	p_3	p_4	
	p_3	p_3	p_5	p_4	p_0	p_2	p_1	
	p_4	p_4	p_3	p_5	p_1	p_0	p_2	
	p_5	p_5	p_4	p_3	p_2	p_1	p_0	

From the nature of the problem we expect to find that this group is isomorphic with the group of the equilateral triangle $\{1, \omega, \omega^2, p, q, r\}$, and this is clearly the case if we establish that 1:1 correspondence $1 \leftrightarrow p_0$, $\omega \leftrightarrow p_1$, $\omega^2 \leftrightarrow p_2$, $p \leftrightarrow p_3$, $q \leftrightarrow p_4$, $r \leftrightarrow p_5$. This permutation group is usually denoted S_3 and is of order 3! or 6. In general the permutation group Sn is of order $n!$.

We notice that S_3 contains subgroups $\{p_0 p_3\}$, $\{p_0 p_4\}$, $\{p_0 p_5\}$, $\{p_0 \, p_1 \, p_2\}$ and no others. We expect these, in any case, for the first three are isomorphic with the group $\{1, q\}$, i.e. the "group of the letter A", and the last is isomorphic with the group of rotations of the equilateral triangle $\{1, \omega, \omega^2\}$.

An interesting thing to note, however, is that in every case we have considered the *order of subgroups has always been a factor of the order of the group*. Thus the group of the square (order 8) had subgroups of order 4 and 2. The group S_3 (order 6) we see has subgroups order 3 and 2. This is a general property.

It is at this point that the serious business of finite group theory begins. The mathematician can determine the groups of symmetries associated with certain problems or mathematical situations; his analysis of these problems, his decision as to whether such problems are reducible to simpler terms and soluble, often depends upon whether those groups of symmetries contain within themselves simpler patterns or subgroups. In some cases it may be possible to show that those groups or their subgroups are isomorphic or homomorphic with groups whose properties are better known. It was by using this type of technique that Abel, using Galois's theory of groups, was able to show that, in general, the polynomial equation of the fifth degree is insoluble except by approximate methods.

We conclude by stating certain results which are useful in this context, although the proofs are beyond the scope of this chapter.

1. The order of a finite group is divisible by the orders of its subgroups.

It follows that

2. If S is a group of order p (prime), then S has no subgroups except itself and the identity element.

3. A group S of prime order p is a cyclic group. (The cyclic group of order n is of the form $\{1, a, a^2, a^3, \ldots, a^{n-1}\}$.)

4. If S is an infinite cyclic group it is isomorphic to the group of integers with respect to the law of addition.

5. If S is a finite cyclic group of order n it is isomorphic to the group of integers modulo n.

6. A group S of order n is isomorphic to a subgroup of the group of permutations Sn.

Useful Reference Books

ADLER, I., *The New Mathematics*, New American Library.
ALEXANDROFF, P. S., *Introduction to the Theory of Groups*, Blackie.
ALLEN, R. G. D., *Basic Mathematics*, Macmillan.
LEDERMANN, W., *Introduction to the Theory of Finite Groups*, Oliver & Boyd.
MANSFIELD, D. E. and BRUCKHEIMER, M., *Background to Set and Group Theory*, Chatto and Windus.
MOAKES, A. J., *The Core of Mathematics*, Macmillan.
PAPY., *Groups*, Macmillan.
REYNOLDS, J. A. C., *Shape, Size and Place*, E. Arnold.
SAWYER, W. W., *Prelude to Mathematics*, Penguin.
SAWYER, W. W., *Concrete Approach to Abstract Algebra*, Freeman.
An Introduction to Groups, Mathematics Teaching Pamphlet No. 11.
Symmetry Groups, Mathematics Teaching Pamphlet No. 12.

4
MATRICES

ROTATIONS IN THE EUCLIDEAN AND COMPLEX PLANES

Transformations are of great importance and wide application in all branches of mathematics. Successive translations or rotations of an invariant shape or rigid body in two-dimensional and three-dimensional space give rise to the important structure which we have called a group. In the last chapter we examined in particular the quarter-turns of the square. There it was sufficient to denote a particular quarter-turn operation by some letter. If, however, we wish to produce a counter-clockwise quarter-turn of some vector $\overrightarrow{OP}$ in the Argand plane, we need some tool or operator which, acting on $\overrightarrow{OP}$, produces this effect. The operator required is i where $i = \sqrt{-1}$.

Thus, if
$$\overrightarrow{OP} = x + iy,$$
$$i \cdot \overrightarrow{OP} = ix - y$$
$$= -y + ix$$
$$= OP'$$

and $\angle POP'$ is clearly one right angle (Fig. 47).

In order to produce a counter-clockwise rotation of ϕ radians we require a different operator.

In polar form
$$\overrightarrow{OP} = r(\cos\theta + i\sin\theta)$$
$$= r\, e^{i\theta}$$

MATRICES 69

Now
$$\overrightarrow{OP''} = r\,e^{i(\theta+\phi)}$$
$$= r\,e^{i\theta} \cdot e^{i\phi}$$
$$= e^{i\phi} \cdot \overrightarrow{OP}.$$

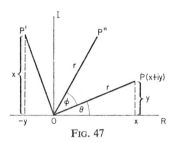

Fig. 47

The operator that rotates $\overrightarrow{OP}$ to $\overrightarrow{OP''}$, or "sends" P to P'' is therefore $e^{i\phi}$. (The quarter-turn is simply a special case, for if $\phi = \tfrac{1}{2}\pi$, $e^{i\phi} = e^{i\frac{1}{2}\pi} = \cos\tfrac{1}{2}\pi + i\sin\tfrac{1}{2}\pi = i$.)

QUARTER-TURNS IN THE CARTESIAN PLANE

In the present chapter we start by considering operators which produce similar effects in the Cartesian plane. In Fig. 48 OP_1, OP_2, OP_3 represent OP_0 after successive quarter-turns. In order

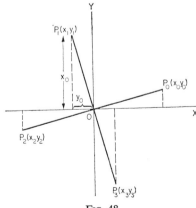

Fig. 48

to send P_0 to a position P_1, it should be clear from Fig. 48 that we make the substitution

$$x_1 = -y_0$$
$$y_1 = x_0$$

or, in full,
$$x_1 = 0x_0 - 1y_0$$
$$y_1 = 1x_0 + 0y_0$$

We now write this pair of equations as follows:

$$\begin{pmatrix} x_1 \\ y_1 \end{pmatrix} = \begin{pmatrix} 0 & -1 \\ 1 & 0 \end{pmatrix} \cdot \begin{pmatrix} x_0 \\ y_0 \end{pmatrix}$$

i.e. we write them in *matrix* form.

A matrix is simply an array of elements placed in rows and columns. When we speak of an "$n \times m$ matrix" we mean a matrix having n rows and m columns. If $n = m$, the matrix is described as a *square matrix*. In the matrix equation above, the operator or transforming matrix $\begin{pmatrix} 0 & -1 \\ 1 & 0 \end{pmatrix}$ is a 2×2 square matrix. It is important to understand that a matrix does not have a *value* in the sense that a determinant has a value. If, however, a matrix $\begin{pmatrix} a & b \\ c & d \end{pmatrix}$ is such that $\begin{vmatrix} a & b \\ c & d \end{vmatrix} = ad - bc = 0$, the matrix is described as a *singular* matrix.

A matrix may consist of a single row such as (a, b, c, d) or a single column such as

$$\begin{pmatrix} x_1 \\ y_1 \end{pmatrix} \quad \text{or} \quad \begin{pmatrix} x_0 \\ y_0 \end{pmatrix} \quad \text{or} \quad \begin{pmatrix} x_2 \\ y_2 \\ z_2 \end{pmatrix}$$

Such matrices are sometimes described as *row vectors* or *column vectors* respectively. In the matrix equation above the point $(x_1 y_1)$ has been written as a column matrix $\begin{pmatrix} x_1 \\ y_1 \end{pmatrix}$, or alternatively we may say that the vector $\overrightarrow{OP_1}$, has been written as a column vector $\begin{pmatrix} x_1 \\ y_1 \end{pmatrix}$.

MATRICES

If we now wish to send the point P_1 to P_2 we must make the substitutions

$$x_2 = -y_1$$
$$y_2 = x_1$$

i.e.
$$\begin{pmatrix} x_2 \\ y_2 \end{pmatrix} = \begin{pmatrix} 0 & -1 \\ 1 & 0 \end{pmatrix} \begin{pmatrix} x_1 \\ y_1 \end{pmatrix}$$

As might be expected, the transforming matrix is the same as before since it produces precisely the same effect as before, i.e. a counter-clockwise quarter-turn. The operator $\begin{pmatrix} 0 & -1 \\ 1 & 0 \end{pmatrix}$ acting in the Cartesian plane corresponds to the operator i in the Argand plane.

Consider now the transformation of the point P_0 to P_2. This can be done in two ways. Transforming directly we have

$$x_2 = -x_0$$
$$y_2 = -y_0$$

i.e.
$$\begin{pmatrix} x_2 \\ y_2 \end{pmatrix} = \begin{pmatrix} -1 & 0 \\ 0 & -1 \end{pmatrix} \cdot \begin{pmatrix} x_0 \\ y_0 \end{pmatrix}$$

or, transforming in two stages via P, we have

$$\begin{pmatrix} x_2 \\ y_2 \end{pmatrix} = \begin{pmatrix} 0 & -1 \\ 1 & 0 \end{pmatrix} \begin{pmatrix} x_1 \\ y_1 \end{pmatrix} \quad \text{and} \quad \begin{pmatrix} x_1 \\ y_1 \end{pmatrix} = \begin{pmatrix} 0 & -1 \\ 1 & 0 \end{pmatrix} \cdot \begin{pmatrix} x_0 \\ y_0 \end{pmatrix}$$

Therefore

$$\begin{pmatrix} x_2 \\ y_2 \end{pmatrix} = \begin{pmatrix} 0 & -1 \\ 1 & 0 \end{pmatrix} \cdot \begin{pmatrix} 0 & -1 \\ 1 & 0 \end{pmatrix} \cdot \begin{pmatrix} x_0 \\ y_0 \end{pmatrix}$$

It follows that the "product"

$$\begin{pmatrix} 0 & -1 \\ 1 & 0 \end{pmatrix} \cdot \begin{pmatrix} 0 & -1 \\ 1 & 0 \end{pmatrix} = \begin{pmatrix} -1 & 0 \\ 0 & -1 \end{pmatrix}$$

What is the rule of "multiplication" which we must adopt in order that this should be so? The clue is provided by considering a more general case.

Let
$$x_2 = ax_1 + by_1 \quad \text{and} \quad x_1 = \alpha x_0 + \beta y_0$$
$$y_2 = cx_1 + dy_1 \qquad\qquad y_1 = \gamma x_0 + \delta y_0$$

By direct substitution these reduce to the single substitution

$$x_2 = (a\alpha + b\gamma)x_0 + (a\beta + b\delta)y_0$$
$$y_2 = (c\alpha + d\gamma)x_0 + (c\beta + d\delta)y_0$$

Writing in matrix form we see that

$$\begin{pmatrix} a & b \\ c & d \end{pmatrix} \cdot \begin{pmatrix} \alpha & \beta \\ \gamma & \delta \end{pmatrix} = \begin{pmatrix} a\alpha + b\gamma & a\beta + \delta b \\ c\alpha + d\gamma & c\beta + \delta d \end{pmatrix}$$

or, briefly, $\quad A \cdot B = C.$

The element in the first row and first column of C is formed by multiplying together corresponding elements in the first row of A and the first column of B, i.e.

The element in the first row and second column of C is formed by multiplying together corresponding elements in the first row of A and the second column of B, i.e.

thus returning to the original example:

$$\begin{pmatrix} 0 & -1 \\ 1 & 0 \end{pmatrix} \cdot \begin{pmatrix} 0 & -1 \\ 1 & 0 \end{pmatrix} = \begin{pmatrix} 0 \times 0 + (-1) \times 1 & 0 \times (-1) + (-1) \times 0 \\ 1 \times 0 + 0 \times 1 & (-1) \times 1 + 0 \times 0 \end{pmatrix}$$

$$= \begin{pmatrix} -1 & 0 \\ 0 & -1 \end{pmatrix} \text{ as required}$$

or, briefly, $\quad L \cdot L = M.$

MATRICES 73

If we now operate upon M with L, that is to say form the product $L.M$, we should produce the transforming matrix required to send P_0 to P_3. This is

$$\begin{pmatrix} 0 & -1 \\ 1 & 0 \end{pmatrix} \cdot \begin{pmatrix} -1 & 0 \\ 0 & -1 \end{pmatrix} = \begin{pmatrix} 0 & 1 \\ -1 & 0 \end{pmatrix} = N, \text{ say,}$$

i.e.
$$x_3 = y_0$$
$$y_3 = -x_0, \text{ which is clearly correct.}$$

Taking one step more, the operation of L on N should produce a transforming matrix which will bring P back on to itself. This is an identity operation and the resulting matrix I is called the *unit matrix*.

Thus:
$$L.N = \begin{pmatrix} 0 & -1 \\ 1 & 0 \end{pmatrix} \begin{pmatrix} 0 & 1 \\ -1 & 0 \end{pmatrix} = \begin{pmatrix} 1 & 0 \\ 0 & 1 \end{pmatrix} = I$$

the unit matrix. Whence L is the *inverse matrix of N* and vice versa. In the last chapter we saw that the operations $\{1, i, -1, -i\}$ formed a commutative group under multiplication. The reader may care to verify, as an exercise, that the matrices I, L, M, N form an isomorphic group under matrix multiplication. As a further exercise the reader may investigate the effects of the matrices

$$\begin{pmatrix} 1 & 0 \\ 0 & 1 \end{pmatrix}, \begin{pmatrix} 1 & 0 \\ 0 & -1 \end{pmatrix}, \begin{pmatrix} -1 & 0 \\ 0 & 1 \end{pmatrix}, \begin{pmatrix} -1 & 0 \\ 0 & -1 \end{pmatrix}$$

and show that these form a group isomorphic with the group of the rectangle $\{I, p, q, r\}$.

GENERAL ROTATION IN THE CARTESIAN PLANE

Imagine a rotation of the Cartesian plane about O in which the axes remain fixed (Fig. 49). Suppose that the point P_0 moves to a position P_1, the vector $\overrightarrow{OP_0}$ turning through ϕ radians. We require the matrix which transforms the column vector $\begin{pmatrix} x_0 \\ y_0 \end{pmatrix}$ to $\begin{pmatrix} x_1 \\ y_1 \end{pmatrix}$.

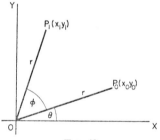

Fig. 49

Now
$$r \cos(\theta+\phi) = r \cos\theta \cos\phi - r \sin\theta \sin\phi,$$
$$\therefore x_1 = x_0 \cos\phi - y_0 \sin\phi$$

and
$$r \sin(\theta+\phi) = r \sin\theta \cos\phi + r \cos\theta \sin\phi$$
$$\therefore y_1 = y_0 \cos\phi + x_0 \sin\phi$$

whence
$$\begin{pmatrix} x_1 \\ y_1 \end{pmatrix} = \begin{pmatrix} \cos\phi & -\sin\phi \\ \sin\phi & \cos\phi \end{pmatrix} \cdot \begin{pmatrix} x_0 \\ y_0 \end{pmatrix}.$$

ENLARGEMENTS

In order to "stretch" the vector $\overrightarrow{OP_0}$ until it becomes the vector $\overrightarrow{OP_1}$ such that $\lambda \cdot OP_0 = OP_1$ (Fig. 50), we require that

$$x_1 = \lambda x_0$$
$$y_1 = \lambda y_0$$

or
$$\begin{pmatrix} x_1 \\ y_1 \end{pmatrix} = \begin{pmatrix} \lambda & 0 \\ 0 & \lambda \end{pmatrix} \cdot \begin{pmatrix} x_0 \\ y_0 \end{pmatrix}.$$

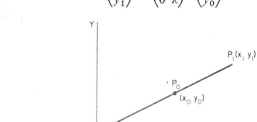

Fig. 50

We write this transformation in the form

$$\lambda \begin{pmatrix} 1 & 0 \\ 0 & 1 \end{pmatrix} \quad \text{or} \quad \lambda I.$$

In general, the multiplication of a matrix by some scalar quantity λ obeys the rule

$$\lambda \begin{pmatrix} p & q \\ r & s \end{pmatrix} = \begin{pmatrix} \lambda p & \lambda q \\ \lambda r & \lambda s \end{pmatrix}.$$

As we see later, this follows from the definition of the rule of addition of matrices.

TRANSFORMATION OF AN AREA

All the transforming matrices mentioned so far have been special cases of the general second order square matrix $\begin{pmatrix} p & q \\ r & s \end{pmatrix}$. Consider the effect of such a transformation upon the vertices $(0, 0), (1, 0), (1, 1), (0, 1)$ of a unit square.

If P_0 is the point $(0, 0)$ $\quad P_1 = \begin{pmatrix} p & q \\ r & s \end{pmatrix}\begin{pmatrix} 0 \\ 0 \end{pmatrix} = \begin{pmatrix} 0 \\ 0 \end{pmatrix}.$

If P is the point $(1, 0)$ $\quad P_1 = \begin{pmatrix} p & q \\ r & s \end{pmatrix}\begin{pmatrix} 1 \\ 0 \end{pmatrix} = \begin{pmatrix} p \\ r \end{pmatrix}.$

For the point $\quad (1, 1) \quad P_1 = \begin{pmatrix} p & q \\ r & s \end{pmatrix}\begin{pmatrix} 1 \\ 1 \end{pmatrix} = \begin{pmatrix} p+q \\ r+s \end{pmatrix}$

and for $\quad\quad\quad (0, 1) \quad P_1 = \begin{pmatrix} p & q \\ r & s \end{pmatrix}\begin{pmatrix} 0 \\ 1 \end{pmatrix} = \begin{pmatrix} q \\ s \end{pmatrix}$

Thus the square, vertices P_0, is transformed into a parallelogram, vertices P_1 (Fig. 51).

The transformation

$$\begin{pmatrix} x_1 \\ y_1 \end{pmatrix} = \begin{pmatrix} p & q \\ r & s \end{pmatrix}\begin{pmatrix} x \\ y \end{pmatrix} \quad \text{or} \quad \begin{cases} x_1 = px+qy \\ y_1 = rx+sy \end{cases}$$

is the general two-dimensional *linear transformation*, and the study of such transformations is an important part of *linear* algebra.

In problems in the study of strain which involve geometrical distortion, we often find that a small element of material of square cross-section when unstrained, becomes distorted under strain into one whose cross-section is a parallelogram (to the first order of

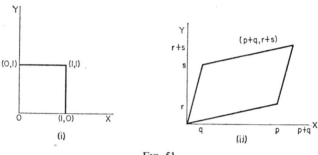

Fig. 51

small quantities). This is the case with rigid materials under irrotational stresses and small vibrations and with liquids or gases flowing irrotationally. The mathematical analysis in such cases involves the use of the linear transformation.

If we apply the transformation to the circle $x^2+y^2 = 1$ we obtain an equation of the second degree, i.e. the circle is transformed into some form of central conic. A special case of a simple kind is provided by applying the transformation

$$\begin{pmatrix} X \\ Y \end{pmatrix} = \begin{pmatrix} 1 & 0 \\ 0 & k \end{pmatrix} \begin{pmatrix} x \\ y \end{pmatrix} \quad \text{to the circle} \quad x^2+y^2 = 1.$$

Now $x^2+y^2 = 1$ may be written in the matrix form

$$(x \ y) \begin{pmatrix} x \\ y \end{pmatrix} = 1;$$

also, the inverse transformation is simply

$$\begin{pmatrix} x \\ y \end{pmatrix} = \begin{pmatrix} 1 & 0 \\ 0 & 1/k \end{pmatrix} \begin{pmatrix} X \\ Y \end{pmatrix}$$

Hence the transformed conic becomes

$$(X,\ Y/k) \begin{pmatrix} 1 & 0 \\ 0 & 1/k \end{pmatrix} \begin{pmatrix} X \\ Y \end{pmatrix} = 1$$

or
$$X^2 + \frac{Y^2}{k^2} = 1,$$

which is an ellipse, semi-axes $1, k$ and eccentricity $\sqrt{1-k^2}$ ($0 < k < 1$). In more difficult cases it is necessary to be able to determine the inverse matrix, a process which we consider later in this chapter.

GENERAL PROPERTIES OF MATRICES

Although new syllabuses in mathematics include the use and manipulation of matrices, it is unlikely that matrices more complex than the second-order square matrix will be dealt with before the sixth-form stage. In the first place such a new concept must be introduced in the simplest possible way in order to avoid obscuring the principle with unnecessarily complicated manipulative techniques. Secondly, many of the results and methods illustrated using second-order square matrices are quite general, and once these have been mastered extension to three-dimensional transformations and harder cases is simply a matter of technique.

However, in order to provide the reader with a broader view of these techniques and some more general results, we include a few more difficult cases in the following paragraphs.

The *general rule of matrix multiplication* states that given a matrix A with l rows and m columns (an $l \times m$ matrix) and an $m \times n$ matrix B, then the product matrix $A.B$ (or C) is an $l \times n$ matrix, such that if c_{rs} is the element in the rth row and sth column of C, it is the sum of the products of elements in the rth row of A with corresponding elements in the sth column of B.

Concisely,
$$c_{rs} = \sum_{t=1}^{m} a_{rt} b_{ts}$$

or, diagrammatically

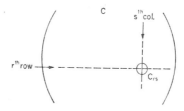

The sum of the products $\sum_{t=1}^{m} a_{rt}b_{ts}$

$= a_{r1}b_{1s} + a_{r2}b_{2s} + a_{r3}b_{3s} \ldots a_{rm}b_{ms}$
$=$ the element in the rth row, sth column of C.

Obviously, the number of *columns* of A must equal the number of *rows* of B, otherwise the product does not exist. The matrix $B.A$, for example, does not exist unless $n = l$, in which case $B.A$ would be an mth order square matrix. Even where $A.B$ and $B.A$ both exist, they are not, in general, equal. Matrix algebra in general is non-commutative under multiplication.

It has already been stressed that a matrix does not possess a value. In a determinant, rows and columns or multiples thereof may be added, subtracted or interchanged without altering its numerical value. These rules do not apply to matrices. If we add together two rows of a matrix we produce a different matrix; if we interchange rows and columns of a matrix A we also produce a different matrix, but in this particular case the result is $\tilde{A}$, the *transposed matrix*. *Two matrices are said to be equal* only if they are of the same order (or dimensions) and if elements in corresponding positions are identical.

Examples

(i) If $A = \begin{pmatrix} 1 & 0 & 2 \\ 0 & 3 & 4 \end{pmatrix}$, $B = \begin{pmatrix} 1 & 2 & 1 & 4 \\ 2 & 0 & 3 & 1 \\ 5 & 4 & 0 & 2 \end{pmatrix}$,

$$A.B = \begin{pmatrix} (1 \times 1 + 0 \times 2 + 2 \times 5)(2 \times 1 + 0 \times 0 + 2 \times 4) \\ (1 \times 1 + 3 \times 0 + 2 \times 0)(1 \times 4 + 0 \times 1 + 2 \times 2) \\ (0 \times 1 + 2 \times 3 + 4 \times 5)(2 \times 0 + 3 \times 0 + 4 \times 4) \\ (0 \times 1 + 3 \times 3 + 4 \times 0)(0 \times 4 + 3 \times 1 + 4 \times 2) \end{pmatrix}$$

$$= \begin{pmatrix} 11 & 10 & 1 & 8 \\ 26 & 16 & 9 & 11 \end{pmatrix}.$$

$B.A$ does not exist.

(ii) If $A = \begin{pmatrix} 2 & 1 \\ 3 & -4 \end{pmatrix}$, $B = \begin{pmatrix} 3 & 0 \\ 1 & 2 \end{pmatrix}$,

$A.B = \begin{pmatrix} 7 & 2 \\ 5 & -8 \end{pmatrix}$, $B.A = \begin{pmatrix} 6 & 3 \\ 8 & -7 \end{pmatrix}$, $A.B \neq B.A$.

(iii) If $A = (2, 3, -1, 0)$, $B = \begin{pmatrix} -1 \\ 2 \\ 4 \\ 1 \end{pmatrix}$,

$A.B = -2+6-4+0 = 0$, $B.A = \begin{pmatrix} -2 & -3 & 1 & 0 \\ 4 & 6 & -2 & 0 \\ 8 & 12 & -4 & 0 \\ 2 & 3 & -1 & 0 \end{pmatrix}.$

(iv) If $A = \begin{pmatrix} 3 & 1 & -2 \\ 6 & 2 & -4 \end{pmatrix}$, $B = \begin{pmatrix} 2 & 1 \\ 4 & 3 \\ 5 & 3 \end{pmatrix}$, $A.B = \begin{pmatrix} 0 & 0 \\ 0 & 0 \end{pmatrix}$

This is called a *null matrix*. Note that $A.B =$ a null matrix does *not* imply that either A or B are null matrices.

(v) If $A = \begin{pmatrix} a & 0 & 0 \\ 0 & b & 0 \\ 0 & 0 & c \end{pmatrix} \quad B = \begin{pmatrix} l & 0 & 0 \\ 0 & m & 0 \\ 0 & 0 & n \end{pmatrix}.$

A and B are called *diagonal matrices*.

$$A.B = \begin{pmatrix} al & 0 & 0 \\ 0 & bm & 0 \\ 0 & 0 & cn \end{pmatrix}, \quad B.A = \begin{pmatrix} al & 0 & 0 \\ 0 & bm & 0 \\ 0 & 0 & cn \end{pmatrix},$$

i.e. $A.B = B.A$. It is generally true that *diagonal matrices are commutative under multiplication*.

(vi) If the rows and columns of a matrix A are interchanged, the resulting matrix is called the *transposed matrix* and is denoted by $\tilde{A}$.

Let $A = \begin{pmatrix} 1 & 2 \\ 3 & 4 \end{pmatrix}, \quad B = \begin{pmatrix} 5 & 2 \\ 1 & 6 \end{pmatrix}.$

Then $\tilde{A} = \begin{pmatrix} 1 & 3 \\ 2 & 4 \end{pmatrix}, \quad \tilde{B} = \begin{pmatrix} 5 & 1 \\ 2 & 6 \end{pmatrix}.$

Note that $A.B = \begin{pmatrix} 7 & 14 \\ 19 & 30 \end{pmatrix}, \quad \widetilde{A.B} = \begin{pmatrix} 7 & 19 \\ 14 & 30 \end{pmatrix},$

and $\tilde{B}.\tilde{A} = \begin{pmatrix} 7 & 19 \\ 14 & 30 \end{pmatrix}.$

Therefore $\widetilde{A.B} = \tilde{B}.\tilde{A}$.

This result also is generally true.

(vii) If $A = \begin{pmatrix} 1 & 2 \\ 3 & 4 \end{pmatrix}, \quad B = \begin{pmatrix} 5 & 2 \\ 1 & 6 \end{pmatrix}.$

Calculate $|A|$, $|B|$, and $|A.B|$

$|A| = 4-6 = -2$

$|B| = 30-2 = 28$

$|A.B| = \begin{vmatrix} 7 & 14 \\ 19 & 30 \end{vmatrix} = 210-266 = -56$

Note that $|A| \cdot |B| = |A.B|$

This again is a result which is true in general.

Addition of matrices is defined for matrices of the same order. The sum of two matrices is the matrix obtained by adding together corresponding elements. Subtraction is defined in similar fashion. For example

$$\begin{pmatrix} 1 & 2 \\ 3 & 4 \end{pmatrix} + \begin{pmatrix} 2 & -1 \\ -2 & 0 \end{pmatrix} = \begin{pmatrix} 3 & 1 \\ 1 & 4 \end{pmatrix}$$

$$\begin{pmatrix} 5 & 7 & 9 \\ 4 & 6 & -2 \end{pmatrix} - \begin{pmatrix} 3 & 7 & 5 \\ 4 & 0 & 2 \end{pmatrix} = \begin{pmatrix} 2 & 0 & 4 \\ 0 & 6 & -4 \end{pmatrix}.$$

Also $\begin{pmatrix} a & b \\ c & d \end{pmatrix} + \begin{pmatrix} a & b \\ c & d \end{pmatrix} = 2\begin{pmatrix} a & b \\ c & d \end{pmatrix}$ or $\begin{pmatrix} 2a & 2b \\ 2c & 2d \end{pmatrix}$,

i.e. for matrices we have the following rule for *multiplying a matrix by a single scalar quantity*

$$\lambda \begin{pmatrix} a & b \\ c & d \end{pmatrix} = \begin{pmatrix} \lambda a & \lambda b \\ \lambda c & \lambda d \end{pmatrix}.$$

Matrices obey the associative law but not the commutative law in respect of multiplication. Under the definition of addition given above, however, they do obey the commutative and associative laws as well as the distributive law of multiplication over addition.

If A, B and C are matrices for which sums and products are defined the laws of matrix algebra may be summarized as follows:

1. $A+B = B+A$, $A.B \neq B.A$.
2. $A+(B+C) = (A+B)+C$, $A.(B.C) = (A.B).C$.
3. $A.(B+C) = A.B+A.C$.
4. There exist "identity elements"

   (i) the null matrix $\begin{pmatrix} 0 & 0 \\ 0 & 0 \end{pmatrix}$ under addition, and

   (ii) the unit matrix $\begin{pmatrix} 1 & 0 \\ 0 & 1 \end{pmatrix}$ or I under multiplication.

5. There exist inverse matrices such that
   (i) $A+(-A)$ = the null matrix,
   (ii) $A^{-1} \times A = A \times A^{-1} = I$ the unit matrix, provided that $|A| \neq 0$, i.e. provided that A is non-singular.

We have here all the ingredients of a group and, indeed, two finite groups of matrices have already been mentioned. As an example of an infinite group it is easily shown that the set of all non-singular square matrices of fixed order n forms an infinite group. Many results true in ordinary algebra are true in matrix algebra, but we must take care. For example, the result of squaring a binomial

$$(a+b)^2 = a^2 + 2ab + b^2$$

becomes $(A+B)^2 = A^2 + A \cdot B + B \cdot A + B^2$

when A, B are matrices.

We now use the rule of addition in two interesting cases. The transformation for a rotation through ϕ radians in the Cartesian plane was given as

$$\begin{pmatrix} x_1 \\ y_1 \end{pmatrix} = \begin{pmatrix} \cos \phi & -\sin \phi \\ \sin \phi & \cos \phi \end{pmatrix} \cdot \begin{pmatrix} x_0 \\ y_0 \end{pmatrix}$$

Expanding by the rule of addition we have

$$\begin{pmatrix} x_1 \\ y_1 \end{pmatrix} = \left[\begin{pmatrix} \cos \phi & 0 \\ 0 & \cos \phi \end{pmatrix} + \begin{pmatrix} 0 & -\sin \phi \\ \sin \phi & 0 \end{pmatrix} \right] \cdot \begin{pmatrix} x_0 \\ y_0 \end{pmatrix}$$
$$= \left[\begin{pmatrix} 1 & 0 \\ 0 & 1 \end{pmatrix} \cos \phi + \begin{pmatrix} 0 & -1 \\ 1 & 0 \end{pmatrix} \sin \phi \right] \cdot \begin{pmatrix} x_0 \\ y_0 \end{pmatrix}$$

Remembering that

$$\begin{pmatrix} 1 & 0 \\ 0 & 1 \end{pmatrix} = I \quad \text{and} \quad \begin{pmatrix} 0 & -1 \\ 1 & 0 \end{pmatrix}$$

corresponded to the operator i in the Argand plane, this transformation becomes strongly reminiscent of

$$OP_1 = (I \cos \phi + i \sin \phi) \cdot OP_0$$

or $e^{i\phi} \cdot OP_0$ which, of course, is the corresponding operator in the Argand plane producing an equal rotation.

SYMMETRIC AND ANTISYMMETRIC MATRICES

A square matrix A, or (a_{rs}), is called *symmetric* if $a_{rs} = a_{sr}$, i.e. if $A = \tilde{A}$. For example

$$A = \begin{pmatrix} 1 & 4 & 5 \\ 4 & 2 & 3 \\ 5 & 3 & 0 \end{pmatrix}, \quad \tilde{A} = \begin{pmatrix} 1 & 4 & 5 \\ 4 & 2 & 3 \\ 5 & 3 & 0 \end{pmatrix} = A$$

If, however, $a_{rs} = -a_{sr}$ or $A = -\tilde{A}$, the matrix A is called *antisymmetric*. In this case it follows that elements in the leading diagonal are all zero. For example

$$B = \begin{pmatrix} 0 & 2 & -3 \\ -2 & 0 & 5 \\ 3 & -5 & 0 \end{pmatrix},$$

$$\tilde{B} = \begin{pmatrix} 0 & -2 & 3 \\ 2 & 0 & -5 \\ -3 & 5 & 0 \end{pmatrix} = -\begin{pmatrix} 0 & 2 & -3 \\ -2 & 0 & 5 \\ 3 & -5 & 0 \end{pmatrix} = -B$$

It is an interesting result, following from the definition of matrix addition, that *any matrix can be written as the sum of two matrices, one symmetric and the other antisymmetric*.

For $\quad A = (a_{rs}) = \tfrac{1}{2}(a_{rs}+a_{sr}) + \tfrac{1}{2}(a_{rs}-a_{sr})$

If r and s are interchanged $(a_{rs}+a_{sr})$ remains unchanged while $(a_{rs}-a_{sr})$ changes sign. Thus $\tfrac{1}{2}(a_{rs}+a_{sr})$ is a symmetric and $\tfrac{1}{2}(a_{rs}-a_{sr})$ an unsymmetric matrix.

Example

$$\text{Let } A = \begin{pmatrix} 6 & 4 & -2 \\ 0 & 10 & 16 \\ 8 & -12 & 14 \end{pmatrix}, \quad \tilde{A} = \begin{pmatrix} 6 & 0 & 8 \\ 4 & 10 & -12 \\ -2 & 16 & 14 \end{pmatrix}$$

$$\tfrac{1}{2}(A+\tilde{A}) = \begin{pmatrix} 6 & 2 & 3 \\ 2 & 10 & 2 \\ 3 & 2 & 14 \end{pmatrix}, \quad \tfrac{1}{2}(A-\tilde{A}) = \begin{pmatrix} 0 & 2 & -5 \\ -2 & 0 & 14 \\ 5 & -14 & 0 \end{pmatrix}$$

therefore $A = \begin{pmatrix} 6 & 4 & -2 \\ 0 & 10 & 16 \\ 8 & -12 & 14 \end{pmatrix}$

$= \begin{pmatrix} 6 & 2 & 3 \\ 2 & 10 & 2 \\ 3 & 2 & 14 \end{pmatrix} + \begin{pmatrix} 0 & 2 & -5 \\ -2 & 0 & 14 \\ 5 & -14 & 0 \end{pmatrix}$

In the next section we consider the solution of simultaneous linear equations using inverse matrices. Where the number of equations is large a computer is required and the solution may be obtained by applying an iterative process to the set of equations written in matrix form. As an example of an iterative process consider the equations

$$x + 3y = 5$$
$$x + y = 6$$

we write
$$x_{r+1} = 6 - y_r$$
$$y_{r+1} = \tfrac{1}{3}(5 - x_r)$$

Taking $x_1 = y_1 = 0$ and making successive substitutions, x_r, y_r tend to the values of x, y satisfying the equations. Thus:

r	1	2	3	4	5	6	7	8
x_r	0	6	4·3	6·3	5·8	6·4	6·3	6·5
y_r	0	1·7	−·3	·2	−·4	−·3	−·5	−·5

and $x = 6·5$, $y = -·5$ are the correct solutions.

If we now have

$$a_{11}x_1 + a_{12}x_2 \ldots a_{1n}x_n = b_1$$
$$a_{21}x_1 + a_{22}x_2 \ldots a_{2n}x_n = b_2$$
$$\vdots$$
$$a_{n1}x_1 + a_{n2}x_2 \ldots a_{nn}x_n = b_n$$

or
$$\begin{pmatrix} a_{11} & a_{12} & \ldots & a_{1n} \\ a_{21} & & & \\ \vdots & & & \\ \vdots & & & \\ a_{n1} & & & \end{pmatrix} \cdot \begin{pmatrix} x_1 \\ x_2 \\ x_3 \\ \vdots \\ x_n \end{pmatrix} = \begin{pmatrix} b_1 \\ b_2 \\ \vdots \\ \vdots \\ b_n \end{pmatrix}$$

or concisely $A.x = B$, the numerical analyst writes the matrix A as the sum of three matrices, a lower diagonal matrix L, a diagonal matrix D and an upper diagonal matrix U thus:

$$\begin{pmatrix} a_{11} & a_{12} & a_{13} & \cdots & a_{1n} \\ a_{21} & a_{22} & & & \\ a_{31} & & & & \\ \vdots & & & & \\ a_{n1} & & & & \end{pmatrix} = \begin{pmatrix} 0 & 0 & 0 & 0 & & 0 \\ a_{21} & 0 & 0 & & & \\ a_{31} & a_{32} & & & & \\ & & a_{43} & \cdot & & \\ & & & \cdot & \cdot & \\ a_{n1} & & & & a_{n,n-1} & 0 \end{pmatrix} (L)$$

$$+ \begin{pmatrix} a_{11} & & & & & \\ & a_{22} & & & \mathbf{0} & \\ & & a_{33} & \cdot & & \\ \mathbf{0} & & & \cdot & \cdot & \\ & & & & \cdot & a_{nn} \end{pmatrix} (D)$$

$$+ \begin{pmatrix} 0 & a_{12} & a_{13} & & & a_{1n} \\ & 0 & a_{23} & & & \\ & & 0 & & & \\ & & & 0 & & \\ & \mathbf{0} & & & \cdot & \\ & & & & & \cdot \\ & & & & & a_{n-1,n} \\ & & & & & 0 \end{pmatrix} (U)$$

i.e. $\quad\quad\quad\quad (L+D+U).x = B.$

For iteration purposes this matrix equation may be written in the various forms

$$Dx_{r+1} = B - (L+U)x_r$$
$$(L+D)x_{r+1} = B - Ux_r$$

from which it is possible to programme a computer to obtain the solution to the required degree of accuracy.

SOLUTION OF LINEAR SIMULTANEOUS EQUATIONS USING THE INVERSE MATRIX

Consider the simultaneous equations

$$2x+y = 4$$
$$x+2y = 5$$

Solving in the normal way we obtain $x = 1$, $y = 2$. Writing these equations in matrix form we have

$$\begin{pmatrix} 2 & 1 \\ 1 & 2 \end{pmatrix} \begin{pmatrix} x \\ y \end{pmatrix} = \begin{pmatrix} 4 \\ 5 \end{pmatrix}$$

or
$$A.x = C$$
$$\therefore x = A^{-1}.C$$

If we can find A^{-1}, the inverse matrix of

$$\begin{pmatrix} 2 & 1 \\ 1 & 2 \end{pmatrix}$$

we can clearly write down the solution straight away. In this section we discuss ways in which this can be done.

Taking the more general case

$$ax+by = p$$
$$cx+dy = q$$

or
$$\begin{pmatrix} a & b \\ c & d \end{pmatrix} \cdot \begin{pmatrix} x \\ y \end{pmatrix} = \begin{pmatrix} p \\ q \end{pmatrix}$$

and solving in the usual way, we have

$$adx+bdy = pd$$
$$cbx+bdy = bq$$

i.e.
$$x = \frac{pd-bq}{ad-cb}$$

and
$$y = \frac{-pc+aq}{ad-cb}$$

or in matrix form

$$\begin{pmatrix} x \\ y \end{pmatrix} = \frac{1}{ad-bc} \begin{pmatrix} d & -b \\ -c & a \end{pmatrix} \cdot \begin{pmatrix} p \\ q \end{pmatrix}$$

i.e. the inverse of

$$\begin{pmatrix} a & b \\ c & d \end{pmatrix} \text{ is } \frac{\begin{pmatrix} d & -b \\ -c & a \end{pmatrix}}{\begin{vmatrix} a & b \\ c & d \end{vmatrix}}$$

provided $ad \neq bc$.

Example

Using this result in the example quoted above

$$\begin{pmatrix} 2 & 1 \\ 1 & 2 \end{pmatrix} \cdot \begin{pmatrix} x \\ y \end{pmatrix} = \begin{pmatrix} 4 \\ 5 \end{pmatrix}$$

$$\therefore \begin{pmatrix} x \\ y \end{pmatrix} = \frac{1}{2^2 - 1^2} \begin{pmatrix} 2 & -1 \\ -1 & 2 \end{pmatrix} \cdot \begin{pmatrix} 4 \\ 5 \end{pmatrix}$$

$$= \tfrac{1}{3} \begin{pmatrix} 3 \\ 6 \end{pmatrix} = \begin{pmatrix} 1 \\ 2 \end{pmatrix}$$

Hence $\qquad x = 1, \quad y = 2$

Example

$$3x + 2y = 1$$
$$2x - 3y = 5$$

$$\therefore \begin{pmatrix} 3 & 2 \\ 2 & -3 \end{pmatrix} \begin{pmatrix} x \\ y \end{pmatrix} = \begin{pmatrix} 1 \\ 5 \end{pmatrix}$$

The inverse of $\begin{pmatrix} 3 & 2 \\ 2 & -3 \end{pmatrix}$ is $-\tfrac{1}{13} \begin{pmatrix} -3 & -2 \\ -2 & 3 \end{pmatrix}$

$$\therefore \begin{pmatrix} x \\ y \end{pmatrix} = -\tfrac{1}{13} \begin{pmatrix} -3 & -2 \\ -2 & 3 \end{pmatrix} \begin{pmatrix} 1 \\ 5 \end{pmatrix}$$

$$= -\tfrac{1}{13} \begin{pmatrix} -13 \\ 13 \end{pmatrix} = \begin{pmatrix} 1 \\ -1 \end{pmatrix}$$

$$\therefore x = 1, \quad y = -1$$

GENERAL METHOD OF FORMING THE INVERSE MATRIX

If $A = (a_{rs})$, form the transposed matrix $\tilde{A} = (a_{sr})$. We now write down the matrix of cofactors A_{sr}, paying careful regard to sign (i.e. in place of the element a_{jk} write down the cofactor of a_{jk}, which is the determinant of $\tilde{A}$ with the jth row and kth column deleted). Calculate the determinant of $A(|A|$ or $\triangle A)$. Then

$$A^{-1} = \frac{A_{sr}}{\triangle A}$$

Example

If $\qquad A = \begin{pmatrix} 1 & 0 & 2 \\ 4 & 3 & 1 \\ 0 & 1 & 5 \end{pmatrix}$ find A^{-1}

Expanding along the first column,

$$\triangle A = 1 \begin{pmatrix} 3 & 1 \\ 1 & 5 \end{pmatrix} - 4 \begin{pmatrix} 0 & 2 \\ 1 & 5 \end{pmatrix} + 0 \begin{pmatrix} 0 & 2 \\ 3 & 1 \end{pmatrix}$$

$$= 14 + 8 + 0 = 22$$

If $(a_{rs}) = \begin{pmatrix} 1 & 0 & 2 \\ 4 & 3 & 1 \\ 0 & 1 & 5 \end{pmatrix}, \quad (a_{sr}) = \begin{pmatrix} 1 & 4 & 0 \\ 0 & 3 & 1 \\ 2 & 1 & 5 \end{pmatrix}$

$$A_{sr} = \begin{pmatrix} \begin{vmatrix} 3 & 1 \\ 1 & 5 \end{vmatrix} & -\begin{vmatrix} 0 & 1 \\ 2 & 5 \end{vmatrix} & \begin{vmatrix} 0 & 3 \\ 2 & 1 \end{vmatrix} \\ -\begin{vmatrix} 4 & 0 \\ 1 & 5 \end{vmatrix} & \begin{vmatrix} 1 & 0 \\ 2 & 5 \end{vmatrix} & -\begin{vmatrix} 1 & 4 \\ 2 & 1 \end{vmatrix} \\ \begin{vmatrix} 4 & 0 \\ 3 & 1 \end{vmatrix} & -\begin{vmatrix} 1 & 0 \\ 0 & 1 \end{vmatrix} & \begin{vmatrix} 1 & 4 \\ 0 & 3 \end{vmatrix} \end{pmatrix} = \begin{pmatrix} 14 & 2 & -6 \\ -20 & 5 & 7 \\ 4 & -1 & 3 \end{pmatrix}$$

$$\therefore A^{-1} = \tfrac{1}{22} \begin{pmatrix} 14 & 2 & -6 \\ -20 & 5 & 7 \\ 4 & -1 & 3 \end{pmatrix}$$

The reader may check that

$$A^{-1}.A = A.A^{-1} = \begin{pmatrix} 1 & 0 & 0 \\ 0 & 1 & 0 \\ 0 & 0 & 1 \end{pmatrix}$$

Example

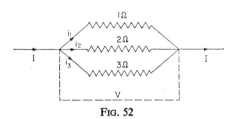

Fig. 52

Calculate i_1, i_2, i_3 and the resistance R of the network in terms of I and Ω (Fig. 52).

Sets of simultaneous linear equations arise in connexion with current networks and the inverse matrix method often provides a quick solution. When the network is complicated and the number of equations is large this is the only method practicable for programming a computer. The circuit shown here is a simple one and the results may be checked using

$$\frac{1}{R} = \frac{1}{\Omega} + \frac{1}{2\Omega} + \frac{1}{3\Omega}, \quad V = IR = i_1\Omega, \text{ etc.}$$

As an illustration of the matrix method we apply Kirchhoff's laws to the network as follows:

$$\begin{aligned} i_1 + i_2 + i_3 &= I \\ i_1 - 2i_2 &= 0 \\ 2i_2 - 3i_3 &= 0 \end{aligned}$$

i.e.
$$\begin{pmatrix} 1 & 1 & 1 \\ 1 & -2 & 0 \\ 0 & 2 & -3 \end{pmatrix} . \begin{pmatrix} i_1 \\ i_2 \\ i_3 \end{pmatrix} = \begin{pmatrix} I \\ 0 \\ 0 \end{pmatrix}$$

If $(a_{rs}) = A = \begin{pmatrix} 1 & 1 & 1 \\ 1 & -2 & 0 \\ 0 & 2 & -3 \end{pmatrix}$, $a_{sr} = \begin{pmatrix} 1 & 1 & 0 \\ 1 & -2 & 2 \\ 1 & 0 & -3 \end{pmatrix}$

$A_{sr} = \begin{pmatrix} 6 & 5 & 2 \\ 3 & -3 & 1 \\ 2 & -2 & -3 \end{pmatrix}$ and $\Delta A = 6-(-3)+2 = 11$

$$\therefore \begin{pmatrix} i_1 \\ i_2 \\ i_3 \end{pmatrix} = \tfrac{1}{11} \begin{pmatrix} 6 & 5 & 2 \\ 3 & -3 & 1 \\ 2 & -2 & -3 \end{pmatrix} \cdot \begin{pmatrix} I \\ 0 \\ 0 \end{pmatrix}$$

$V = IR = i_1 \Omega$, $\therefore R = \dfrac{i_1 \Omega}{I} = \dfrac{6\Omega}{11}$

It is not suggested that the matrix method is the best in this particular case. We have deliberately selected an example which can be solved in other ways.

Example

Solve the equations

$$x+y+z = 6 \\ 2x-y+3z = 9 \\ 3x+2y-2z = 1$$

In matrix form we have

$$\begin{pmatrix} 1 & 1 & 1 \\ 2 & -1 & 3 \\ 3 & 2 & -2 \end{pmatrix} \cdot \begin{pmatrix} x \\ y \\ z \end{pmatrix} = \begin{pmatrix} 6 \\ 9 \\ 1 \end{pmatrix}$$

If A or $(a_{rs}) = \begin{pmatrix} 1 & 1 & 1 \\ 2 & -1 & 3 \\ 3 & 2 & -2 \end{pmatrix}$, $a_{sr} = \begin{pmatrix} 1 & 2 & 3 \\ 1 & -1 & 2 \\ 1 & 3 & -2 \end{pmatrix}$

$\Delta A = -4+13+7 = 16$, $A_{sr} = \begin{pmatrix} -4 & 4 & 4 \\ 13 & -5 & -1 \\ 7 & 1 & -3 \end{pmatrix}$

$$\therefore \begin{pmatrix} x \\ y \\ z \end{pmatrix} = \tfrac{1}{16} \begin{pmatrix} -4 & 4 & 4 \\ 13 & -5 & -1 \\ 7 & 1 & -3 \end{pmatrix} \cdot \begin{pmatrix} 6 \\ 9 \\ 1 \end{pmatrix} = \tfrac{1}{16} \begin{pmatrix} 16 \\ 32 \\ 48 \end{pmatrix} = \begin{pmatrix} 1 \\ 2 \\ 3 \end{pmatrix}$$

$$\therefore x = 1, \quad y = 2, \quad z = 3$$

AN ALTERNATIVE METHOD OF DETERMINING THE INVERSE MATRIX

The matrix method of solving simultaneous linear equations as presented so far will appeal to the sixth-form mathematician whose knowledge of determinants may enable him to understand the general method of forming the inverse matrix; at any rate he will be able to discover for himself the general rule as it applies to second-order square matrices. As a means of introducing an application of matrices to junior children the method is open to serious objections.

It is unlikely that 12- or 13-year-olds will be able to "discover" the inverse of the second-order matrix, and even if they can be led to see that the inverse of

$$\begin{pmatrix} a & b \\ c & d \end{pmatrix} \text{ is } \frac{1}{ad-bc} \begin{pmatrix} d & -b \\ -c & a \end{pmatrix}$$

it may well seem rather like a new trick, another formula to be remembered. This is not to say, however, that the idea of a matrix is beyond such children. Indeed, it is already on the new O-level syllabuses of four examination boards; rightly so, too, since the fundamental ideas of "transformation", "function", "operators" are well illustrated by matrices.

Consider therefore the following alternative approach. In the usual method of solution one unknown is eliminated by multiplying the equations by suitable factors and adding or subtracting the resulting equations. In other words, the essential rules of solution consist of row operations. Any rows may be multiplied by a constant factor other than zero; rows may be added or subtracted. The essential features of

$$2x + y = 4$$
$$x + 2y = 5$$

are the coefficients and constant terms which may be set out in the following array:

$$\begin{array}{cc|c} 2 & 1 & 4 \\ 1 & 2 & 5 \end{array}$$

We perform row operations until the array on the left is

$$\begin{array}{cc|} 1 & 0 \\ 0 & 1 \end{array}$$

(i.e. the unit matrix from which the result is read off).
Thus:

$$\begin{array}{cc|c} 2 & 1 & 4 \\ 1 & 2 & 5 \end{array} \xrightarrow{\text{multiply row 2 by 2}} \begin{array}{cc|c} 2 & 1 & 4 \\ 2 & 4 & 10 \end{array} \xrightarrow{\text{subtract row 1 from row 2}}$$

$$\begin{array}{cc|c} 2 & 1 & 4 \\ 0 & 3 & 6 \end{array} \xrightarrow{\text{divide row 2 by 3}} \begin{array}{cc|c} 2 & 1 & 4 \\ 0 & 1 & 2 \end{array} \xrightarrow{\text{subtract row 2 from row 1}}$$

$$\begin{array}{cc|c} 2 & 0 & 2 \\ 0 & 1 & 2 \end{array} \xrightarrow{\text{divide row 1 by 2}} \begin{array}{cc|c} 1 & 0 & 1 \\ 0 & 1 & 2 \end{array}$$

whence
$$1x + 0y = 1$$
$$0x + 1y = 2,$$
i.e.
$$x = 1, \quad y = 2.$$

There are, of course, many ways of arriving at the final array, but with practice the number of steps can be minimized.

We now look at the same process in greater detail.

$$\begin{pmatrix} 2 & 1 \\ 1 & 2 \end{pmatrix} \cdot \begin{pmatrix} x \\ y \end{pmatrix} = \begin{pmatrix} 1 & 0 \\ 0 & 1 \end{pmatrix} \cdot \begin{pmatrix} 4 \\ 5 \end{pmatrix}$$
$$\downarrow$$
$$\begin{pmatrix} 2 & 1 \\ 2 & 4 \end{pmatrix}, \qquad \begin{pmatrix} 1 & 0 \\ 0 & 2 \end{pmatrix}$$
$$\downarrow$$
$$\begin{pmatrix} 2 & 1 \\ 0 & 3 \end{pmatrix}, \qquad \begin{pmatrix} 1 & 0 \\ -1 & 2 \end{pmatrix}$$
$$\downarrow$$

$$\begin{pmatrix} 2 & 1 \\ 0 & 1 \end{pmatrix}, \qquad \begin{pmatrix} 1 & 0 \\ -\frac{1}{3} & \frac{2}{3} \end{pmatrix}$$
$\downarrow$
$$\begin{pmatrix} 2 & 0 \\ 0 & 1 \end{pmatrix}, \qquad \begin{pmatrix} \frac{4}{3} & -\frac{2}{3} \\ -\frac{1}{3} & \frac{2}{3} \end{pmatrix}$$
$\downarrow$
$$\begin{pmatrix} 1 & 0 \\ 0 & 1 \end{pmatrix} \cdot \begin{pmatrix} x \\ y \end{pmatrix} = \begin{pmatrix} \frac{2}{3} & -\frac{1}{3} \\ -\frac{1}{3} & \frac{2}{3} \end{pmatrix} \cdot \begin{pmatrix} 4 \\ 5 \end{pmatrix}$$
$$\begin{pmatrix} x & 0 \\ 0 & y \end{pmatrix} = \begin{pmatrix} 1 \\ 2 \end{pmatrix}$$
$$\therefore x = 1, \quad y = 2$$

Notice that

$$\begin{pmatrix} \frac{2}{3} & -\frac{1}{3} \\ -\frac{1}{3} & \frac{2}{3} \end{pmatrix} \quad \text{or} \quad \tfrac{1}{3} \begin{pmatrix} 2 & -1 \\ -1 & 2 \end{pmatrix}$$

was in the inverse of $\begin{pmatrix} 2 & 1 \\ 1 & 2 \end{pmatrix}$ as obtained by the previous method.

This clearly suggests an alternative way of finding an inverse matrix. Given a matrix A we write $A \cdot \begin{pmatrix} 1 & 0 \\ 0 & 1 \end{pmatrix}$, or $A \cdot I$ where I is the unit matrix whose order is that of A. We then, by row operations, transform A into the unit matrix. If at each step we carry out the identical row operations on I, then when A has been reduced to I, the matrix I will have become A^{-1}, the inverse of A.

Example

Find the inverse of $\begin{pmatrix} 1 & 1 & 1 \\ 1 & -2 & 0 \\ 0 & 2 & -3 \end{pmatrix}$,

$$\begin{pmatrix} 1 & 1 & 1 \\ 1 & -2 & 0 \\ 0 & 2 & -3 \end{pmatrix} \cdot \begin{pmatrix} 1 & 0 & 0 \\ 0 & 1 & 0 \\ 0 & 0 & 1 \end{pmatrix}$$

$\downarrow$ row 2 − row 1

$$\begin{pmatrix} 1 & 1 & 1 \\ 0 & -3 & -1 \\ 0 & 2 & -3 \end{pmatrix} \cdot \begin{pmatrix} 1 & 0 & 0 \\ -1 & 1 & 0 \\ 0 & 0 & 1 \end{pmatrix}$$

$\downarrow$ row 2 + row 3

$$\begin{pmatrix} 1 & 1 & 1 \\ 0 & -1 & -4 \\ 0 & 2 & -3 \end{pmatrix} \cdot \begin{pmatrix} 1 & 0 & 0 \\ -1 & 1 & 1 \\ 0 & 0 & 1 \end{pmatrix}$$

$\downarrow$ row 3 + 2 row 2

$$\begin{pmatrix} 1 & 1 & 1 \\ 0 & -1 & -4 \\ 0 & 0 & -11 \end{pmatrix} \cdot \begin{pmatrix} 1 & 0 & 0 \\ -1 & 1 & 1 \\ -2 & 2 & 3 \end{pmatrix}$$

$\downarrow$ divide row 3 by -11

$$\begin{pmatrix} 1 & 1 & 1 \\ 0 & -1 & -4 \\ 0 & 0 & 1 \end{pmatrix} \cdot \begin{pmatrix} 1 & 0 & 0 \\ -1 & 1 & 1 \\ \frac{2}{11} & -\frac{2}{11} & -\frac{3}{11} \end{pmatrix}$$

$\downarrow$ row 1 + row 2

$$\begin{pmatrix} 1 & 0 & -3 \\ 0 & -1 & -4 \\ 0 & 0 & 1 \end{pmatrix} \cdot \begin{pmatrix} 0 & 1 & 1 \\ -1 & 1 & 1 \\ \frac{2}{11} & -\frac{2}{11} & -\frac{3}{11} \end{pmatrix}$$

$\downarrow$ row 1 + 3 row 3

$$\begin{pmatrix} 1 & 0 & 0 \\ 0 & -1 & -4 \\ 0 & 0 & 1 \end{pmatrix} \cdot \begin{pmatrix} \frac{6}{11} & \frac{5}{11} & \frac{1}{11} \\ -1 & 1 & 1 \\ \frac{2}{11} & -\frac{2}{11} & -\frac{3}{11} \end{pmatrix}$$

$\downarrow$ row 2 + 4 row 3

$$\begin{pmatrix} 1 & 0 & 0 \\ 0 & -1 & 0 \\ 0 & 0 & 1 \end{pmatrix} \cdot \begin{pmatrix} \frac{6}{11} & \frac{5}{11} & \frac{2}{11} \\ -\frac{3}{11} & \frac{3}{11} & -\frac{1}{11} \\ \frac{2}{11} & -\frac{2}{11} & -\frac{3}{11} \end{pmatrix}$$

$\downarrow$ divide row 2 by -1

$$\begin{pmatrix} 1 & 0 & 0 \\ 0 & 1 & 0 \\ 0 & 0 & 1 \end{pmatrix} \cdot \begin{pmatrix} \frac{6}{11} & \frac{5}{11} & \frac{2}{11} \\ \frac{3}{11} & -\frac{3}{11} & \frac{1}{11} \\ \frac{2}{11} & -\frac{2}{11} & -\frac{3}{11} \end{pmatrix}$$

$\therefore$ the inverse required

$$\frac{1}{11} \begin{pmatrix} 6 & 5 & 2 \\ 3 & -3 & 1 \\ 2 & -2 & -3 \end{pmatrix}$$

(see example on current network).

MATRIX MULTIPLICATION IN ARITHMETICAL PROBLEMS

Certain types of problem in arithmetic lend themselves to matrix methods of multiplication.

A builder develops a site by building 9 houses and 6 bungalows. On the average, 1 house requires 1600 units of materials and 2000 hours of labour, and one bungalow requires 1500 units of materials and 1800 hours of labour. Labour costs 10s. per hour and each unit of material costs on the average 20s. Find the cost of developing the site.

The quickest way to evaluate the total cost is to write the costing analysis in the form of a 2×2 matrix as follows:

	Labour	*Material*
House	2000	1600
Bungalow	1800	1500

the number of buildings as a row vector (9, 6) and the cost per hour per unit as a column vector $\begin{pmatrix} \frac{1}{2} \\ 1 \end{pmatrix}$. The triple matrix product then gives the total cost of the project, i.e.

$$\text{total cost in £} = (9, \ 6) \cdot \begin{pmatrix} 2000 & 1600 \\ 1800 & 1500 \end{pmatrix} \cdot \begin{pmatrix} \frac{1}{2} \\ 1 \end{pmatrix}$$
$$= (18000 + 10800 \ , \ 14400 + 9000) \cdot \begin{pmatrix} \frac{1}{2} \\ 1 \end{pmatrix}$$
$$= 14,400 + 23,400 = 37,800.$$

∴ total cost = £37,800.

Such problems might well provide a starting point for the introduction of matrices to junior children.

EIGENVECTORS AND THE CHARACTERISTIC EQUATION OF A MATRIX

We conclude this chapter with an elementary discussion of the latent roots of a matrix. Work of this type will be left until the sixth-form stage. At that stage, however, it is worth mentioning owing to its great importance in the more advanced applications of matrix theory in geometry and mechanics.

In general, when $X' = AX$, the vector X is changed in direction by the transformation.

Thus, if $\begin{pmatrix} x \\ y \end{pmatrix} = \begin{pmatrix} 1 \\ 1 \end{pmatrix}$ and $A = \begin{pmatrix} 2 & 1 \\ 3 & -1 \end{pmatrix}$

then $\begin{pmatrix} x_1 \\ y_1 \end{pmatrix} = \begin{pmatrix} 2 & 1 \\ 3 & -1 \end{pmatrix} \cdot \begin{pmatrix} 1 \\ 1 \end{pmatrix} = \begin{pmatrix} 3 \\ 2 \end{pmatrix}$

.e. $P_0 \to P_1$ and the vector $\overrightarrow{OP_0}$ is changed in direction to become $\overrightarrow{OP_1}$ (Fig. 53).

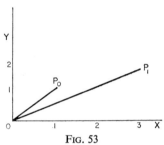

FIG. 53

It happens in some cases, however, that under the transformation by some matrix A, there exists a vector $\begin{pmatrix} x \\ y \end{pmatrix}$ which simply becomes $\begin{pmatrix} \lambda x \\ \lambda y \end{pmatrix}$ or $\lambda \begin{pmatrix} x \\ y \end{pmatrix}$, i.e. the effect of the transformation A upon this particular vector is to change its length (or modulus) in some fixed ratio $\lambda : 1$, but to leave its direction (or argument) unaltered. Such a vector (or vectors) is called a *proper vector* or *eigenvector* of the matrix A.

In such a case we have

$$AX = \lambda X$$

(where A is some matrix, X a column vector, λ a constant),

$$\therefore (A - \lambda I)X = 0,$$
$$\therefore |(A - \lambda I)| = 0.$$

MATRICES

This equation in λ has roots $\lambda_1, \lambda_2 \ldots \lambda_n$. If these are real the corresponding real values of X may be determined. The values λ_r are known as the *eigenvalues* or *latent roots* of the matrix A.

Example

Determine the latent roots of the matrix $\begin{pmatrix} 2 & 1 \\ 1 & 2 \end{pmatrix}$ and the corresponding eigenvectors.

Let $\begin{pmatrix} x \\ y \end{pmatrix}$ be one of the required eigenvectors, λ a constant.

Then
$$\begin{pmatrix} 2 & 1 \\ 1 & 2 \end{pmatrix} \cdot \begin{pmatrix} x \\ y \end{pmatrix} = \lambda \begin{pmatrix} x \\ y \end{pmatrix}$$

$$\therefore \begin{pmatrix} 2 & 1 \\ 1 & 2 \end{pmatrix} \cdot \begin{pmatrix} x \\ y \end{pmatrix} = \lambda \begin{pmatrix} 1 & 0 \\ 0 & 1 \end{pmatrix} \cdot \begin{pmatrix} x \\ y \end{pmatrix} = \begin{pmatrix} \lambda & 0 \\ 0 & \lambda \end{pmatrix} \cdot \begin{pmatrix} x \\ y \end{pmatrix}$$

$$\therefore \begin{vmatrix} 2-\lambda & 1 \\ 1 & 2-\lambda \end{vmatrix} = 0$$

$$\therefore 4 - 4\lambda + \lambda^2 - 1 = 0$$

$$\therefore \lambda^2 - 4\lambda + 3 = 0,$$

i.e. $\lambda = 3$ or $\lambda = 1$

These are the latent roots of $\begin{pmatrix} 2 & 1 \\ 1 & 2 \end{pmatrix}$

If $\lambda = 1$ $\quad \begin{pmatrix} 2 & 1 \\ 1 & 2 \end{pmatrix} \cdot \begin{pmatrix} x \\ y \end{pmatrix} = \begin{pmatrix} x \\ y \end{pmatrix}$

$$\therefore 2x + y = x$$
$$x + 2y = y$$

$$\therefore \frac{x}{y} = -1$$

$\therefore \begin{pmatrix} x \\ y \end{pmatrix}$ could be $\begin{pmatrix} 1 \\ -1 \end{pmatrix}$. The transformation sends this to $\begin{pmatrix} 1 \\ -1 \end{pmatrix}$ i.e. to its original position. Alternatively $\begin{pmatrix} x \\ y \end{pmatrix} = \begin{pmatrix} -1 \\ 1 \end{pmatrix}$ and again this point is also mapped on to itself by the transformation. In general any vector $\begin{pmatrix} k \\ -k \end{pmatrix}$ is mapped on to itself. (All real k.)

If $\lambda = 3$ $\quad \begin{pmatrix} 2 & 1 \\ 1 & 2 \end{pmatrix} \cdot \begin{pmatrix} x \\ y \end{pmatrix} = \begin{pmatrix} 3x \\ 3y \end{pmatrix}$

$$\therefore 2x+y = 3x$$
$$x+2y = 3y$$
$$\therefore \frac{y}{x} = 1,$$

$\therefore$ for example, $\begin{pmatrix} 2 & 1 \\ 1 & 2 \end{pmatrix} \cdot \begin{pmatrix} 2 \\ 2 \end{pmatrix} \rightarrow \begin{pmatrix} 6 \\ 6 \end{pmatrix}$

and any vector $\begin{pmatrix} k \\ k \end{pmatrix}$ is mapped on to $\begin{pmatrix} 3k \\ 3k \end{pmatrix}$

Useful Reference Books

ADLER, I., *The New Mathematics*, New American Library.
AITKEN, A. C., *Determinants and Matrices*, Oliver & Boyd.
ALLEN, R. G. D., *Basic Mathematics*, Macmillan.
ARCHBOLD, J. W., *Algebra*, Pitman.
BALFOUR, A., *An Introduction to Sets, Groups and Matrices*, Heinemann.
COHN, P. M., *Linear Equations*, Routledge & Kegan Paul.
FLETCHER, T. J., *Some Lessons in Mathematics*, C.U.P.
GEARY, A., LOWRY, H. V. and HAYDEN, H. A., *Advanced Mathematics for Technical Students*, Part II, Chapter 6, Longmans.
GIBBS, W. J., *Electric Machine Analysis Using Matrices*, Pitman.
GIBSON, G. R. and MAYATT, J., *First Stages in Matrices*, U.L.P.
HADLEY, G., *Linear Algebra*, Addison-Wesley.
KEMENY, J. G., MIRKIL, H., SNELL, J. L. and THOMPSON, G. L., *Finite Mathematical Structures*, Prentice-Hall.
MATTHEWS, G., *Matrices I and II*, E. Arnold.
MATTHEWS, G., Matrices for the million, *Mathematical Gazette*, Vol. XLVII, No. 359 (Feb. 1963), Bell.
SAWYER, W. W., *Prelude to Mathematics*, Penguin.

5
VECTORS

FREE VECTORS

Most children at some stage during their secondary career are introduced to the idea of a vector. Displacement, velocity, acceleration, momentum and force are all quantities in which not only the magnitude but the direction (and in the case of force the actual line of action) must form part of a complete specification. It is sufficient to say that the temperature of a body is 400°C or that the kinetic energy of a moving truck is 4 foot tons; it is not sufficient to say that in order to get from Sheffield to Leeds I must displace myself 33 miles. Unless I displace myself in the correct direction I may well arrive in Stockport or the outskirts of Nottingham instead.

Quantities in which the magnitude and the direction are essential features are called vector quantities. (Strictly speaking such quantities must also obey the parallelogram law. This is so in all the cases quoted above but there are exceptions. For example, the successive rotations of a rigid body in space through finite angles possess magnitude and direction, but their resultant cannot be obtained by the parallelogram law.)

In the past, the idea of a vector quantity has frequently been introduced with reference to forces acting at a point. Experimentally it is easy to verify that the resultant of two non-parallel forces may be obtained by the parallelogram law, but unfortunately this approach has tended to produce confusion over the idea of a vector.

Force is a quantity in which the line of action as well as the magnitude and direction must be specified and it is therefore a representation of a special type of vector sometimes called a *sliding vector*.

The magnitude and direction of a magnetic field on the other hand must be specified with respect to some given point. This is an example of the representation of another special vector localized at a point and therefore termed a *tied vector*.

Finally we have quantities which are completely specified by magnitude and direction alone. The moment of a couple, for example, is independent of the point or axis about which moments are taken. It is a representation of what is appropriately termed a *free vector*. In what follows, unless otherwise stated, we shall be dealing with free vectors.

In order to emphasize and clarify the idea of a free vector consider a rigid body which is displaced without rotation a certain distance in some given direction (Fig. 54).

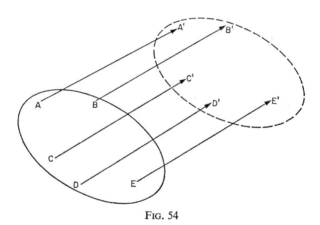

Fig. 54

Particles A, B, C, D, E move to new positions A', B', C', D', E', and the arrows show how each point moves. The point is that any one of them is sufficient to describe the displacement of the body. Denoting the displacement from A to A' as $\overrightarrow{AA'}$ we see that $\overrightarrow{AA'}$ and $\overrightarrow{BB'}$ are not only equal, they are *equivalent*. For the truth

VECTORS 101

is that *a vector is a translation of the whole space* and $\overrightarrow{AA'}$, $\overrightarrow{BB'}$, etc. are equivalent representations of it. If we decide to call $\overrightarrow{AA'}$ and $\overrightarrow{BB'}$ vectors, then they are the same vector. Put in more sophisticated terms, *a vector is an equivalence class of displacements* and not a displacement itself.

[A similar situation occurs in arithmetic. We cannot add the fractions $\frac{2}{3}$ and $\frac{1}{2}$ but we can add the rational numbers they represent; for the rational number $\frac{2}{3}$ is the equivalence class $\frac{2x}{3x}$ for all x ($\neq 0$) and we can add the representatives $\frac{4}{6}$ and $\frac{3}{6}$.]

Having disentangled the notion of a vector from that of a physical vector quantity we can proceed more freely to develop a vector approach to school geometry. As we shall see, this approach is often powerful and elegant and frequently yields exciting new solutions with no more knowledge required than the laws of addition and scalar multiplication of two vectors. Moreover, vectors provide yet another example of a non-commutative algebra and an infinite group under addition.

At the university level, dynamics and statics, hydrodynamics and electromagnetics are usually developed in vector terms, and in geometry a thorough understanding of the concept of a vector space is a basic requirement. In this chapter we discuss what might be done to develop the stage A and stage B approach to this work.

REPRESENTATION AND ADDITION OF VECTORS

Vectors will be denoted by small letters **a**, **b**, **c** in bold type or by letters with arrows above $\overrightarrow{PQ}$, $\overrightarrow{QR}$, etc. In written work the vector **a** is distinguished from the scalar a by a short underline thus $\underline{a}$. A vector **a** may be represented in magnitude and direction by a line segment $\overrightarrow{PQ}$. Since the exact position is not important any other parallel and equal segment $\overrightarrow{SR}$ also represents the vector **a** The magnitude of the vector **a** is denoted by $|\mathbf{a}|$ and this is read as the *modulus of* **a**. A *unit vector* is one whose modulus is unity. A vector whose modulus is zero is called a *null* vector. Two vectors **a** and **b** are said to be *equal* if they have equal moduli and act in the same direction. For this we write **a** = **b**. The vector equal in magnitude and opposite in direction to **a** is written as −**a**, and if **a** is represented by $\overrightarrow{PQ}$ then −**a** is represented by $\overrightarrow{QP}$.

(i) Consider the parallelogram *PQRS* (Fig. 55). Let $\overrightarrow{PQ}$, $\overrightarrow{QR}$ represent the vectors **a**, **b** respectively.

Then $$\overrightarrow{PQ} = \overrightarrow{SR} = \mathbf{a}; \quad \overrightarrow{QR} = \overrightarrow{PS} = \mathbf{b}$$

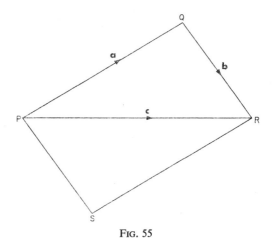

FIG. 55

If $\overrightarrow{PQ}$ represents the displacement of a particle from *P* to *Q* and $\overrightarrow{QR}$ the displacement from *Q* to *R*, then the effect of making these displacements in succession is the same as the effect of the direct displacement $\overrightarrow{PR}$.

i.e. $$\overrightarrow{PQ} + \overrightarrow{QR} = PR$$
or $$\mathbf{a} + \mathbf{b} = \mathbf{c}$$

This is the familiar triangle rule of vector addition. We may also write

$$\vec{PS} + \vec{SR} = PR$$

or
$$\mathbf{b} + \mathbf{a} = \mathbf{c}$$

Whence
$$\mathbf{a} + \mathbf{b} = \mathbf{b} + \mathbf{a},$$

i.e. the commutative law of addition holds.

(ii) Let $\vec{AD}$, $\vec{DC}$, $\vec{CB}$ (Fig. 56) represent the vectors **p, q, r**.

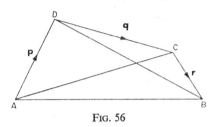

Fig. 56

Considering successive displacements $\vec{AD}$, $\vec{DB}$ we have

$$\vec{AB} = \vec{AD} + \vec{DB} = \vec{AD} + (\vec{DC} + \vec{CB})$$

i.e.
$$\vec{AB} = \mathbf{p} + (\mathbf{q} + \mathbf{r})$$

while for successive displacements $\vec{AC}$, $\vec{CB}$ we have

$$\vec{AB} = \vec{AC} + \vec{CB} = (\vec{AD} + \vec{DC}) + \vec{CB}$$

$$\therefore \vec{AB} = (\mathbf{p} + \mathbf{q}) + \mathbf{r}$$

Whence $\mathbf{p} + (\mathbf{q} + \mathbf{r}) = (\mathbf{p} + \mathbf{q}) + \mathbf{r}$ and the associative law of addition also holds.

(iii) If k is a scalar quantity or pure number and **l** is a vector, then we define $k\mathbf{l}$ as a vector parallel to the vector **l** but of modulus k times as great.

Thus the triangles ABC, $A'B'C'$ are similar and $\overrightarrow{A'C'}$ is parallel to $\overrightarrow{AC}$ (Fig. 57).

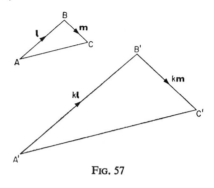

Fig. 57

Now
$$\overrightarrow{AC} = \mathbf{l} + \mathbf{m},$$
$$\therefore \overrightarrow{A'C'} = k(\mathbf{l} + \mathbf{m}) = k\mathbf{l} + k\mathbf{m}.$$

Thus scalar multiplication is distributive over vector addition.

POSITION VECTORS AND MATRICES

In this section we shall see that the triangle law of addition of vectors and the commutative, associative and distributive properties all follow immediately from the rules of addition and scalar multiplication of column matrices or vectors as given in the last chapter. We could in fact make this our starting point, particularly if the topics were developed in this way in the classroom. If, however, elementary work with vectors is commenced before the need to define addition of matrices is encountered then the approach suggested in the previous section may be preferable. In any case it is vitally important at some stage to emphasize the close connexion between matrices and vectors.

In Fig. 58 (x_1, y_1) are the coordinates which specify the position of P relative to the rectangular axes OX, OY. (We have taken rectangular axes for simplicity, but of course the same results apply in the case of oblique axes. The point is worth making with more mature pupils since it emphasizes the fact that the addition and numerical multiplication of vectors have purely geometrical interpretations that do not depend on the particular coordinate system.)

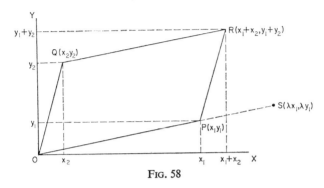

FIG. 58

We can equally well specify the position of P relative to O by $\overrightarrow{OP}$ which is called the *position vector of P relative to O*. (Clearly there exists a one:one correspondence between the set of points P of the plane and the set of arrows $\overrightarrow{OP}$ which can be drawn from O to P.)

In the last chapter we denoted $\overrightarrow{OP}$ by the column vector or column matrix $\begin{pmatrix} x_1 \\ y_1 \end{pmatrix}$. We also defined the sum of two matrices of like order. As a special case column matrices or column vectors are added according to the law

$$\begin{pmatrix} x_1 \\ y_1 \end{pmatrix} + \begin{pmatrix} x_2 \\ y_2 \end{pmatrix} = \begin{pmatrix} x_1 + x_2 \\ y_1 + y_2 \end{pmatrix}$$

i.e. $\begin{pmatrix} x_1 + x_2 \\ y_1 + y_2 \end{pmatrix}$ is the column vector sum of $\overrightarrow{OP}$ and $\overrightarrow{OQ}$.

From Fig. 58 we see that the gradients and lengths of $\overrightarrow{OQ}$ and $\overrightarrow{PR}$ are equal, whence OPRQ is a parallelogram. In other words the law of addition of column matrices implies the parallelogram law of addition of vectors

$$\vec{OP} + \vec{OQ} = \vec{OR}.$$

[This is to be expected since the writing of $\vec{OP}$ as a column vector $\begin{pmatrix} x_1 \\ y_1 \end{pmatrix}$ in the first place implies the parallelogram law, but this will not be obvious to children.]

In the last chapter we also defined the scalar product of a matrix as

$$\lambda \begin{pmatrix} a & b \\ c & d \end{pmatrix} = \begin{pmatrix} \lambda a & \lambda b \\ \lambda c & \lambda d \end{pmatrix}$$

In the case of a column matrix this gives

$$\lambda \begin{pmatrix} x_1 \\ y_1 \end{pmatrix} = \begin{pmatrix} \lambda x_1 \\ \lambda y_1 \end{pmatrix}$$

or $\qquad \lambda \vec{OP} = \vec{OS}, \quad \text{where} \quad |\vec{OS}| = \lambda |\vec{OP}|.$

Therefore O, P, S are collinear points and the vectors represented by $\vec{OP}, \vec{OS}$ have the same direction. This again is consistent with the definition given in the last section for the product of k with the free vector $\mathbf{l}$.

It is important to realize that position vectors are free vectors. The position of Q relative to O is the same as the position of R relative to P. Alternatively, a vector is a translation of the whole space and the displacements $\vec{OQ}$ and $\vec{PR}$ are equivalent representations of it. In either case $\vec{OQ} = \vec{PR}$ and $\vec{OP} = \vec{QR}$. If this were not so then, of course, the commutative law of vector addition would fail.

A single free vector cannot by itself represent the effect of a vector quantity such as force which is tied to a line. We shall see later that in such a case two free vectors are necessary.

APPLICATIONS TO GEOMETRY

(i) *The midpoint theorem.* This follows immediately from the definitions already given. For if (Fig. 59) X, Y are midpoints of AB, AC respectively we have

$$\overrightarrow{XA} + \overrightarrow{AY} = \overrightarrow{XY}$$

$$\therefore 2\overrightarrow{XA} + 2\overrightarrow{AY} = 2\overrightarrow{XY}$$

i.e. $$\overrightarrow{BA} + \overrightarrow{AC} = 2\overrightarrow{XY}$$

but $$\overrightarrow{BA} + \overrightarrow{AC} = \overrightarrow{BC}$$

$$\therefore \overrightarrow{BC} = 2\overrightarrow{XY}$$

Hence $\overrightarrow{BC}$ is parallel to $\overrightarrow{XY}$ and $BC = 2XY$.

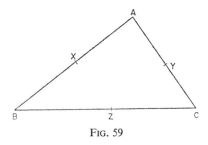

FIG. 59

(ii) *Division of a line in a given ratio.* We are given fixed points P, Q with position vectors $\overrightarrow{OP}$, $\overrightarrow{OQ}$ relative to O (Fig. 60). We wish to find the position vector of a point R which divides PQ in the ratio $l:m$.

Now $$m\overrightarrow{OP} + m\overrightarrow{PR} = m\overrightarrow{OR} \qquad (1)$$

and $$l\overrightarrow{OQ} + l\overrightarrow{QR} = l\overrightarrow{OR} \qquad (2)$$

If $\dfrac{PR}{RQ} = \dfrac{l}{m}$ then $m\vec{PR} = l\vec{RQ} = -l\vec{QR}$

∴ adding (1) and (2) we have

$$m\vec{OP} + l\vec{OQ} = (m+l)\vec{OR}$$

or
$$\vec{OR} = \frac{m\vec{OP} + l\vec{OQ}}{m+l}$$

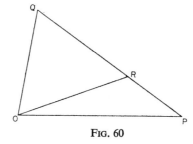

FIG. 60

This result may almost be regarded as a basic theorem in vector geometry. It is of very great use as we shall see. Changing the notation and rewriting the above

$$m\mathbf{p} + l\mathbf{q} - (m+l)\mathbf{r} = 0$$

or
$$\begin{cases} m\mathbf{p} + l\mathbf{q} + n\mathbf{r} = 0 \\ \text{where} \quad m+l+n = 0 \end{cases}$$

gives us the conditions that points P, Q, R (with position vectors $\mathbf{p}, \mathbf{q}, \mathbf{r}$) should be *collinear* (Fig. 61).

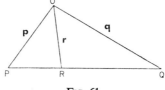

FIG. 61

VECTORS

If vectors $\overrightarrow{OP}, \overrightarrow{OQ}, \overrightarrow{OS}$ are coplanar and P, Q, S are not collinear (Fig. 62), where, say,

$$\lambda\mathbf{r} = \mathbf{s}(\lambda \neq 1)$$

then
$$m\mathbf{p} + l\mathbf{q} = (m+l)\mathbf{r}$$

becomes
$$m\mathbf{p} + l\mathbf{q} = \frac{m+l}{\lambda}\mathbf{s}$$

Writing
$$k = \frac{-(m+l)}{\lambda}$$

we have
$$m\mathbf{p} + l\mathbf{q} + k\mathbf{s} = 0 \qquad (1)$$

but
$$m + l + k \neq 0$$

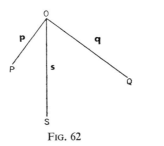

FIG. 62

(1) is the condition that the vectors **p**, **q**, **s** are coplanar.

[We sometimes say that if $m\mathbf{p} + n\mathbf{q} + l\mathbf{r} = 0$ for m, n, l all non-zero, then the vectors **p**, **q**, **r** are *linearly dependent*.]

(iii) *The medians of a triangle.* Let the position vectors of points A, B, C relative to O be **a**, **b**, **c**. Let X, Y be midpoints of AB, BC respectively (Fig. 63).

We have proved in (i) that $\overrightarrow{AC} = 2\overrightarrow{XY}$. It follows that triangles XYG, CAG are similar if $\dfrac{XG}{GC} = \dfrac{1}{2}$. By the theorem proved in (ii) the position vector of X is

$$\frac{\mathbf{a}+\mathbf{b}}{2}$$

and hence the position of vector of G is

$$\frac{2\overrightarrow{OX}+\overrightarrow{OC}}{2+1} \quad \text{or} \quad \frac{2\left(\dfrac{\mathbf{a}+\mathbf{b}}{2}\right)+\mathbf{c}}{3} \quad \text{or} \quad \frac{\mathbf{a}+\mathbf{b}+\mathbf{c}}{3} \tag{1}$$

If we replace $\mathbf{b}$ by $\mathbf{c}$ and $\mathbf{c}$ by $\mathbf{b}$, i.e. apply the permutation

$$\begin{pmatrix} a & b & c \\ a & c & b \end{pmatrix}$$

we obtain the position vector of a point which divides the line $\overrightarrow{ZB}$ in the ratio 1:2, where Z is the midpoint of AC. That is to say we

obtain $$\frac{\mathbf{a}+\mathbf{c}+\mathbf{b}}{3}$$

which is equal to (1), the position vector of G. Hence the medians of $\triangle ABC$ are concurrent through G.

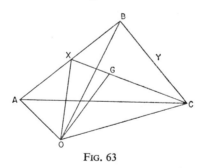

Fig. 63

The result (1) that $3\overrightarrow{OG} = \overrightarrow{OA}+\overrightarrow{OB}+\overrightarrow{OC}$ is a special case of the general result that if G is the centroid of n points $A_1, A_2, \ldots, A_n$ then $n\overrightarrow{OG} = \overrightarrow{OA_1}+\overrightarrow{OA_2} \ldots \overrightarrow{OA_n}$. The reader may care to verify this.

(iv) *The angle bisector theorem.*

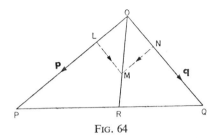

Fig. 64

Let the position vectors of P, Q, relative to O be **p**, **q** respectively (Fig. 64). If OR bisects $\angle POQ$, NM is parallel to OP. LM is parallel to OQ, then $OLMN$ is a rhombus and $OL = ON$. Suppose now that OL, ON are of unit length then $\vec{OL}$, $\vec{ON}$ are unit

vectors and
$$\vec{OM} = \vec{OL} + \vec{LM}$$
$$= \vec{OL} + \vec{ON}$$

∴ for all values of λ, **r** is a point on the bisector of $\angle POQ$, where

$$\mathbf{r} = \lambda(\vec{OL} + \vec{ON})$$

Now $\qquad OP . \vec{OL} = \mathbf{p} \quad \text{and} \quad OQ . \vec{ON} = \mathbf{q}$

$$\therefore \mathbf{r} = \lambda\left(\frac{\mathbf{p}}{OP} + \frac{\mathbf{q}}{OQ}\right) \quad \text{or} \quad \mathbf{r} = \lambda\left(\frac{OQ\mathbf{p} + OP\mathbf{q}}{OP . OQ}\right)$$

The last three expressions for **r** are, of course, vector equations of the bisector OR.

If we now take $\qquad \lambda = \dfrac{OP . OQ}{OQ + OP}$

then $\qquad \mathbf{r} = \dfrac{OQ . \mathbf{p} + OP . \mathbf{q}}{OQ + OP}$

But this, by example (ii), is the condition that

$$\frac{PR}{RQ} = \frac{OP}{OQ}$$

Hence the bisector of any angle of a triangle divides the side opposite to that angle in the ratio of the sides containing the bisector.

RESOLUTION OF A VECTOR

If we introduce unit vectors **i**, **j** along the rectangular axes OX, OY (Fig. 65), we may write

$$\overrightarrow{ON} = \mathbf{x}_1 = x_1\mathbf{i}$$
$$\overrightarrow{NP} = \mathbf{y}_1 = y_1\mathbf{j}$$

$\overrightarrow{ON}$ and $\overrightarrow{NP}$ are the resolutes of the vector $\overrightarrow{OP}$ along the axes OX, OY so that

$$\overrightarrow{OP} = \overrightarrow{ON} + \overrightarrow{NP}$$

i.e. $\quad\mathbf{r}_1 = x_1\mathbf{i} + y_1\mathbf{j}$

Similarly, $\quad\mathbf{r}_2 = x_2\mathbf{i} + y_2\mathbf{j}$

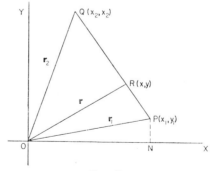

Fig. 65

These equations are very useful and we shall use them again later. Meanwhile

$$\vec{PQ} = \mathbf{r}_2 - \mathbf{r}_1$$
$$= (x_2 - x_1)\mathbf{i} + (y_2 - y_1)\mathbf{j}$$

∴ the gradient of $PQ = \dfrac{y_2 - y_1}{x_2 - x_1}$

and $\qquad PQ = \sqrt{(x_2 - x_1)^2 + (y_2 - y_1)^2}$

If, as in example (ii) of the previous section,

$$\frac{PR}{RQ} = \frac{l}{m},$$

then
$$\vec{OR} = \frac{m\vec{OP} + l\vec{OQ}}{m + l}$$

or
$$\mathbf{r} = \frac{m\mathbf{r}_1 + l\mathbf{r}_2}{m + l}$$

In terms of the coordinates

$$x\mathbf{i} + y\mathbf{j} = \frac{m}{m+l}(x_1\mathbf{i} + y_1\mathbf{j}) + \frac{l}{m+l}(x_2\mathbf{i} + y_2\mathbf{j})$$

Comparing coefficients of **i** and **j** we have

$$x = \frac{mx_1 + lx_2}{m+l}, \quad y = \frac{my_1 + ly_2}{m+l}$$

another familiar result in coordinate geometry.

If R is *any* point on PQ, then

$$\mathbf{r} = \mathbf{r}_1 + \lambda \vec{PQ}$$

or $\qquad \mathbf{r} = \mathbf{r}_1 + \lambda(\mathbf{r}_2 - \mathbf{r}_1)$

or $\qquad \mathbf{r} = (1 - \lambda)\mathbf{r}_1 + \lambda \mathbf{r}_2$

or $\qquad \mathbf{r} = t_1\mathbf{r}_1 + t_2\mathbf{r}_2, \quad \text{where } t_1 + t_2 = 1$

This is the vector equation of the line joining the points with position vectors $\mathbf{r}_1$, $\mathbf{r}_2$.

From this we may deduce the Cartesian equation of the line joining the points $(x_1 y_1)(x_2 y_2)$. Writing the vector equation of the line PQ in terms of the coordinates,

$$x\mathbf{i}+y\mathbf{j} = t_1(x_1\mathbf{i}+y_1\mathbf{j})+t_2(x_2\mathbf{i}+y_2\mathbf{j})$$

Comparing coefficients of $\mathbf{i}$ and $\mathbf{j}$ and using $t_1 = 1-t_2$ we have

$$x = x_1+t_2(x_2-x_1)$$
$$y = y_1+t_2(y_2-y_1)$$

Eliminating t_2, the equation of PQ becomes

$$\frac{y-y_1}{y_2-y_1} = \frac{x-x_1}{x_2-x_1}$$

THE SCALAR PRODUCTS OF TWO VECTORS

The scalar product of two vectors $\mathbf{a}$ and $\mathbf{b}$ is defined as the product of their moduli multiplied by the cosine of the angle θ between their directions. It is written

$$\mathbf{a}.\mathbf{b} = ab \cos\theta = \mathbf{b}.\mathbf{a}.$$

For the scalar product $\mathbf{a}.\mathbf{a}$,

$$\theta = 0, \cos\theta = 1 \quad \text{and} \quad \mathbf{a}.\mathbf{a} = a^2$$

If $\mathbf{a}$ and $\mathbf{b}$ are perpendicular,

$$\theta = \tfrac{1}{2}\pi \quad \text{and} \quad \cos\tfrac{1}{2}\pi = 0$$
$$\therefore \mathbf{a}.\mathbf{b} = 0$$

Or, conversely, if the scalar product of two vectors is zero, the vectors are perpendicular.

For the unit vectors i and j we have

$$\mathbf{i}.\mathbf{j} = 0, \quad \mathbf{i}.\mathbf{i} = \mathbf{j}.\mathbf{j} = 1$$

The scalar product of a vector over vector addition is distributive.

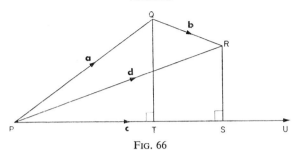

Fig. 66

Let $\overrightarrow{PQ}$, $\overrightarrow{QR}$, $\overrightarrow{PR}$, $\overrightarrow{PU}$ represent the vectors **a**, **b**, **d**, **c** respectively (Fig. 66).

Then
$$(\mathbf{a}+\mathbf{b}).\mathbf{c} = \mathbf{d}.\mathbf{c} = PS.PU$$
$$\mathbf{a}.\mathbf{c} = PT.PU$$
$$\mathbf{b}.\mathbf{c} = TS.PU$$
$$\therefore \mathbf{a}.\mathbf{c}+\mathbf{b}.\mathbf{c} = PU.(PT+TS) = PU.PS$$

Hence $(\mathbf{a}+\mathbf{b}).\mathbf{c} = \mathbf{a}.\mathbf{c}+\mathbf{b}.\mathbf{c}$

Having established the distributive law we can now show that the definition of the scalar product is designed to be consistent with the usual definition of the cosine, for in Fig. 67

$$\mathbf{c} = \mathbf{a}+\mathbf{b} \qquad (1)$$

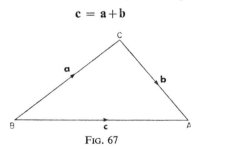

Fig. 67

Multiply scalarly by **c** when

$$\mathbf{c}.\mathbf{c} = \mathbf{a}.\mathbf{c}+\mathbf{b}.\mathbf{c}$$

i.e. $c^2 = ac \cos B + bc \cos A$

or $c = a \cos B + b \cos A$

a standard result in trigonometry.

Again, if we square (scalarly) both sides of equation (1)

$$\mathbf{c}.\mathbf{c} = \mathbf{a}.\mathbf{a} + \mathbf{b}.\mathbf{b} + 2\mathbf{a}.\mathbf{b}$$
$$\therefore c^2 = a^2 + b^2 + 2ab \cos(\pi - c)$$
or $$c^2 = a^2 + b^2 - 2ab \cos C$$

which is the cosine rule.

In the particular case, where $\angle C = 90°$, $\mathbf{a}.\mathbf{b} = 0$,

$$\therefore c^2 = a^2 + b^2$$

which is Pythagoras' theorem.

Reverting to the vectors $\mathbf{r}_1, \mathbf{r}_2$ of the last section and forming their scalar product we find that

$$\mathbf{r}_1.\mathbf{r}_2 = (x_1\mathbf{i} + y_1\mathbf{j}).(x_2\mathbf{i} + y_2\mathbf{j})$$
$$= x_1 x_2 + y_1 y_2$$

But this is the matrix product $(x_1 y_1)\begin{pmatrix} x_2 \\ y_2 \end{pmatrix}$, i.e. the scalar product of two vectors is the matrix product (or inner product) of the components of the vectors.

In three dimensions we introduce a third unit vector $\mathbf{k}$ along the Z-axis so that we now have

$$\mathbf{i}.\mathbf{j} = \mathbf{j}.\mathbf{k} = \mathbf{k}.\mathbf{i} = 0; \quad \mathbf{i}.\mathbf{i} = \mathbf{j}.\mathbf{j} = \mathbf{k}.\mathbf{k} = 1$$

For two vectors
$$\mathbf{r}_1 = x_1\mathbf{i} + y_1\mathbf{j} + z_1\mathbf{k}$$
$$\mathbf{r}_2 = x_2\mathbf{i} + y_2\mathbf{j} + z_2\mathbf{k}$$

we still have $\mathbf{r}_1.\mathbf{r}_2$, the matrix product

$$(x_1, \quad y_1, \quad z_1) \begin{pmatrix} x_2 \\ y_2 \\ z_2 \end{pmatrix}$$

Thus we see that while the matrix law of addition implied the vector law of addition, the definition of the matrix product implies our definition of the scalar product of two vectors.

APPLICATIONS OF THE SCALAR PRODUCT

We now give six examples of geometrical proofs and problems worked by vector methods including the use of scalar multiplication. Work of this kind provides good practice in the elementary rules and an interesting way of looking at familiar results.

One should not overlook the fact that our vector definitions imply many results of Euclidean geometry. Once we have defined the length of a vector we have built in all the metrical properties of the Euclidean plane. The definition that if $n\mathbf{a} = m\mathbf{b}$ then $\mathbf{a}$ is parallel to $\mathbf{b}$, is another way of assuming the truth of Euclid's postulate of parallelism and all results which are based on that assumption.

The vector method is not always the quickest or most elegant one available. Just as there are problems which are better treated by the methods of coordinate geometry, so we have problems which are better treated by vector methods. The point is that vectors provide another tool for the mathematician and it is only through use and practice that he comes to recognize the best situations in which to use it.

Example (i)

If a line is drawn from the centre of a circle to the midpoint of a chord, the line is perpendicular to the chord (Fig. 68).

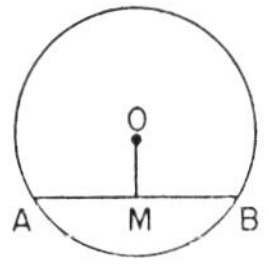

Fig. 68

Proof

$$\overrightarrow{OM} + \overrightarrow{MA} = \overrightarrow{OA} \qquad (1)$$

$$\overrightarrow{OM} + \overrightarrow{MB} = \overrightarrow{OB}$$

i.e. $$\overrightarrow{OM} - \overrightarrow{MA} = \overrightarrow{OB} \qquad (2)$$

Squaring (1) and (2) scalarly, and subtracting, we have

$$4\overrightarrow{OM}.\overrightarrow{MA} = 0$$

∴ $\overrightarrow{OM}$ is perpendicular to $\overrightarrow{AB}$.

Example (ii)

The angle in a semicircle is a right angle (Fig. 69).

Fig. 69

Proof

$$\overrightarrow{CA} = \overrightarrow{CO} + \overrightarrow{OA}$$

$$\overrightarrow{CB} = \overrightarrow{CO} + \overrightarrow{OB}$$

$$= \overrightarrow{CO} - \overrightarrow{OA}$$

$$\therefore \overrightarrow{CA}.\overrightarrow{CB} = CO^2 - OA^2 = 0$$

∴ $\overrightarrow{CA}$ is perpendicular to $\overrightarrow{CB}$.

Example (iii)

$A(1,2)$; $B(5,-3)$; $C(4,3)$; $D(2,6)$ are the vertices of a quadrilateral (Fig. 70). Calculate $\angle DAC$ and show that the diagonals intersect at right angles.

Taking unit vectors **i, j** in the directions $\overrightarrow{OX}$, $\overrightarrow{OY}$ respectively we have

$$\overrightarrow{OA} = \mathbf{i}+2\mathbf{j}, \quad \overrightarrow{OB} = 5\mathbf{i}-3\mathbf{j}, \quad \overrightarrow{OC} = 4\mathbf{i}+3\mathbf{j}, \quad \overrightarrow{OD} = 2\mathbf{i}+6\mathbf{j}$$

$$\therefore \overrightarrow{AD} = \overrightarrow{OD} - \overrightarrow{OA} = \mathbf{i}+4\mathbf{j} \quad \text{and} \quad \overrightarrow{AC} = \overrightarrow{OC} - \overrightarrow{OA} = 3\mathbf{i}+\mathbf{j}$$

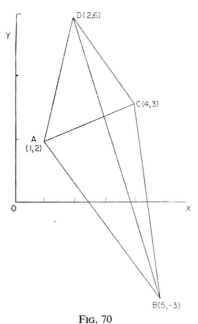

Fig. 70

Now $\vec{AD}.\vec{AC} = |\vec{AC}|.|\vec{AB}| \cos \angle DAC$

i.e. $(\mathbf{i}+4\mathbf{j}).(3\mathbf{i}+\mathbf{j}) = \sqrt{1^2+4^2}\sqrt{3^2+1^2} \cos \angle DAC$

$\therefore 7 = \sqrt{17}\sqrt{10} \cos \angle DAC$

$\therefore \angle DAC = \cos^{-1} \dfrac{7}{\sqrt{170}}$

Further $\vec{BD} = \vec{OD} - \vec{OB} = -3\mathbf{i}+9\mathbf{j}$

$\therefore \vec{AC}.\vec{BD} = (3\mathbf{i}+\mathbf{j}).(-3\mathbf{i}+9\mathbf{j})$

$= -9+9 = 0$

$\therefore \vec{AC}$ is perpendicular to $\vec{BD}$.

An alternative method of setting out is as follows:

Expressing as column vectors we have

$$\vec{OA} = \begin{pmatrix} 1 \\ 2 \end{pmatrix}, \quad \vec{OB} = \begin{pmatrix} 5 \\ -3 \end{pmatrix}, \quad \vec{OC} = \begin{pmatrix} 4 \\ 3 \end{pmatrix}, \quad \vec{OD} = \begin{pmatrix} 2 \\ 6 \end{pmatrix}$$

$$\therefore \vec{AD} = \vec{OD} - \vec{OA} = \begin{pmatrix} 2 \\ 6 \end{pmatrix} - \begin{pmatrix} 1 \\ 2 \end{pmatrix} = \begin{pmatrix} 1 \\ 4 \end{pmatrix}$$

and $\quad \vec{AC} = \vec{OC} - \vec{OA} = \begin{pmatrix} 4 \\ 3 \end{pmatrix} - \begin{pmatrix} 2 \\ 1 \end{pmatrix} = \begin{pmatrix} 3 \\ 1 \end{pmatrix}$

But $\quad \vec{AD} \cdot \vec{AC} = (1, 4) \cdot \begin{pmatrix} 3 \\ 1 \end{pmatrix} = \sqrt{17} \cdot \sqrt{10} \cos \angle DAC$

and $\quad \angle DAC = \cos^{-1} \dfrac{7}{\sqrt{170}} \quad$ as before

Further $\quad \vec{AC} \cdot \vec{BD} = (3, 1) \cdot \begin{pmatrix} -3 \\ 9 \end{pmatrix} = 0$

The inner matrix product vanishes, hence the vectors $\vec{AC}, \vec{BD}$ are perpendicular.

Example (iv)

In Fig. 71

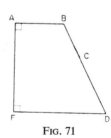

Fig. 71

$AB = BC$ and $CD = DE$. Prove that $\angle ACE = 90°$.

Proof. Take unit vectors **i**, **n** in directions $\vec{AB}, \vec{BC}$ and let $\vec{AB}, \vec{BC}$ have modulus k, $\vec{CD}, \vec{ED}$ modulus n, then

$$\vec{AC} = k(\mathbf{i} + \mathbf{n}), \quad \vec{EC} = n(\mathbf{i} - \mathbf{n})$$

$$\therefore \vec{AC}, \vec{EC} = nk(1 - 1) = 0 \quad \therefore \angle ACE = 90°$$

Exercise

In triangle PQR, $PQ = PR$. S lies on QR and T on PR so that $QS = SR$ and $\angle STR = 90°$. U is the midpoint of ST. Prove that PU is perpendicular to QT.

Vector methods are very useful in three-dimensional problems. Most of the coordinate geometry of three dimensions may be developed from this standpoint.

Example (v)

$ABCD$ with is a tetrahedron AB perpendicular to CD, AD perpendicular to BC (Fig. 72). Prove that AC is also perpendicular to BD.

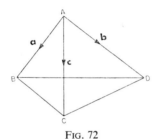

Fig. 72

Proof. Let $\overrightarrow{AB}$, $\overrightarrow{AC}$, $\overrightarrow{AD}$ represent the vectors **a**, **c**, **b** respectively.

$$\overrightarrow{CD} = \mathbf{b} - \mathbf{c}$$

$$\overrightarrow{BC} = \mathbf{c} - \mathbf{a}$$

$$\therefore \mathbf{a}.(\mathbf{b}-\mathbf{c}) = 0 \quad \text{and} \quad \mathbf{b}.(\mathbf{c}-\mathbf{a}) = 0$$

Adding $\qquad \mathbf{b}.\mathbf{c} - \mathbf{a}.\mathbf{c} = 0$

$$\therefore \mathbf{c}.(\mathbf{b}-\mathbf{a}) = 0$$

But $\qquad \mathbf{b} - \mathbf{a} = \overrightarrow{BD}$

$$\therefore \overrightarrow{AC}.\overrightarrow{BD} = 0$$

i.e. AC is perpendicular to BD.

Example (vi)

Find the angle of inclination of any two diagonals of a cube. The angle is the same for any size of cube. Consider, therefore, a unit cube and represent the conterminous edges $\overrightarrow{OB}, \overrightarrow{OE}, \overrightarrow{OD}$ by the usual unit vectors **i, j, k** (Fig. 73).

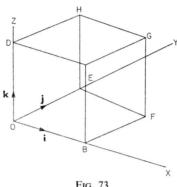

Fig. 73

Then
$$\overrightarrow{OG} = \mathbf{i}+\mathbf{j}+\mathbf{k}$$

and
$$\overrightarrow{DF} = \mathbf{i}+\mathbf{j}-\mathbf{k}$$

$$\therefore \overrightarrow{OG}.\overrightarrow{DF} = (\mathbf{i}+\mathbf{j}+\mathbf{k}).(\mathbf{i}+\mathbf{j}-\mathbf{k})$$
$$= \mathbf{i}^2+\mathbf{j}^2-\mathbf{k}^2+2\mathbf{i}.\mathbf{j}$$
$$= 1+1-1+0$$
$$= 1$$

But $\overrightarrow{OG}.\overrightarrow{DF} = |OG|.|DF|\cos\phi$, where ϕ is the acute angle between the diagonals

$$= \sqrt{3}.\sqrt{3}\cos\phi$$
$$\therefore \phi = \cos^{-1}\tfrac{1}{3}.$$

THE VECTOR PRODUCT OF TWO VECTORS

The scalar product of two vectors is a pure number. For other purposes it is convenient to define a further product, the vector product (or cross product) which is a *vector*. The rule of combination is as follows.

If two vectors **a** and **b** are inclined at an angle θ, we define the vector product $\mathbf{a} \times \mathbf{b}$ of the two vectors as that vector which has magnitude $ab \sin \theta$ and direction perpendicular to the plane containing **a** and **b** in the sense of a right hand screw from **a** to **b**. Thus in Fig. 74

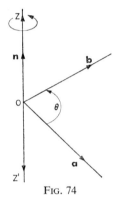

FIG. 74

$$\mathbf{a} \times \mathbf{b} = \overrightarrow{ab \cdot \sin \theta} \quad \text{in the direction } \overrightarrow{OZ}$$

and $$\mathbf{b} \times \mathbf{a} = \overrightarrow{ab \cdot \sin \theta} \quad \text{in the direction } \overrightarrow{OZ'}$$

If **n** is a unit vector in the direction $\overrightarrow{OZ}$

$$\mathbf{a} \times \mathbf{b} = -\mathbf{b} \times \mathbf{a} = ab \sin \theta \, \mathbf{n}$$

In the special case $\theta = \frac{1}{2}\pi$, $\mathbf{a} \times \mathbf{b} = ab\mathbf{n}$, whereas if $\theta = 0$, $\mathbf{a} \times \mathbf{b} = 0$.

There is an analogy here with matrix multiplication. In the first place, as with matrix products, vector products are not commutative. Further, just as the product of two non-null matrices may be a null matrix, so the product of two non-zero vectors may be zero. However, it can be proved that vectors obey the distributive law of vector multiplication over addition.

For the mutually perpendicular unit vectors **i**, **j**, **k** the various vector products are

$$\mathbf{i}^2 = \mathbf{j}^2 = \mathbf{k}^2 = 0$$
$$\mathbf{i} \times \mathbf{j} = \mathbf{k}, \quad \mathbf{j} \times \mathbf{k} = \mathbf{i}, \quad \mathbf{k} \times \mathbf{i} = \mathbf{j}$$

VECTOR AREA

We now introduce the idea of a plane area S having associated with it the direction of the unit vector **n** which acts along the normal to S. In the case where S is a parallelogram whose

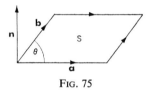

Fig. 75

adjacent sides are represented by vectors **a**, **b** inclined at angle θ (Fig. 75), we see that

$$\mathbf{a} \times \mathbf{b} = ab \cdot \sin\theta \cdot \mathbf{n} = S\mathbf{n} = \mathbf{S}$$

i.e. $S\mathbf{n}$ or **S** is the vector area of S.

APPLICATIONS OF THE VECTOR PRODUCT

Consider a force acting on a particle. If the particle is displaced in a direction which is not perpendicular to the line of action of the force, then the latter does work during the displacement. If **F**, **d** are vectors representing the force and displacement respectively, inclined at an angle θ, then in appropriate units the work done is measured by $Fd \cos\theta$ or $\mathbf{F} \cdot \mathbf{d}$.

We note that the vector **F** is a free vector. It represents the force above in magnitude and direction only. In order to specify the line of action another free vector is required. Now we can specify a force completely if, in addition to its magnitude and direction, we also give its moment about a specified point, say, O.

Let **r** be the position vector relative to O of any point P on the line of action of the force (Fig. 76). Then the moment of the force about O is represented by a vector of magnitude $r \sin\theta \cdot F$, i.e.

taking into account the sense, the moment is represented by the vector $\mathbf{r} \times \mathbf{F}$. The vectors $\mathbf{F}$ and $\mathbf{r} \times \mathbf{F}$ together completely specify the force.

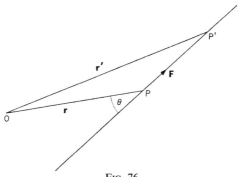

Fig. 76

Further, we note that the moment of the force may also be represented by $\mathbf{r}' \times \mathbf{F}$. But

$$\mathbf{r}' \times \mathbf{F} = (\mathbf{r} + \overrightarrow{PP'}) \times \mathbf{F} = \mathbf{r} \times \mathbf{F}$$

since

$$\overrightarrow{PP'} \times \mathbf{F} = 0.$$

Hence the moment of the force is independent of the position of the point P on its line of action.

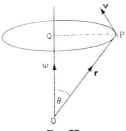

Fig. 77

As an application in dynamics, consider a point P position vector $\mathbf{r}$ relative to 0 at some instant of time, and a vector $\boldsymbol{\omega}$ which represents the magnitude, direction and sense of the angular velocity of the point P (Fig. 77). Then the linear velocity of P at this instant of time is represented by a vector **v** of magnitude $\omega \cdot QP$ or $\omega r \sin \theta$ acting in a direction perpendicular to the plane QOP. But this is precisely the vector given by $\boldsymbol{\omega} \times \mathbf{r}$: i.e. $\mathbf{v} = \boldsymbol{\omega} \times \mathbf{r}$.

We may use this result to deduce the components v_x, v_y, v_z of the velocity of P in space relative to fixed axes OX, OY, OZ respectively. In terms of the unit vectors **i, j, k**,

$$\boldsymbol{\omega} = \omega_x \mathbf{i} + \omega_y \mathbf{j} + \omega_z \mathbf{k}$$
$$\mathbf{r} = x\mathbf{i} + y\mathbf{j} + z\mathbf{k}$$

Hence, remembering that $\mathbf{i}^2 = \mathbf{j}^2 = \mathbf{k}^2 = 0$, $\mathbf{i} \times \mathbf{j} = \mathbf{k}$, etc.,

$$\boldsymbol{\omega} \times \mathbf{r} = \mathbf{i}(z\omega_y - y\omega_z) + \mathbf{j}(x\omega_z - z\omega_x) + \mathbf{k}(y\omega_x - x\omega_y)$$

or

$$\begin{vmatrix} \mathbf{i} & \mathbf{j} & \mathbf{k} \\ \omega_x & \omega_y & \omega_z \\ x & y & z \end{vmatrix}$$

Since $v = v_x i + v_y j + v_z k$, comparing coefficients of the unit vectors we have

$$v_x = z\omega_y - y\omega_z$$
$$v_y = x\omega_z - z\omega_x$$
$$v_z = y\omega_x - x\omega_y$$

APPLICATIONS TO GEOMETRY

Example (i)

The area of a triangle.

The vector area of the triangle OPQ (Fig. 78)

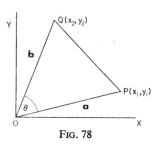

Fig. 78

$$= \tfrac{1}{2}\mathbf{a} \times \mathbf{b} = \tfrac{1}{2}(x_1\mathbf{i} + y_1\mathbf{j}) \times (x_2\mathbf{i} + y_2\mathbf{j})$$
$$= \tfrac{1}{2}(x_1 y_2 - x_2 y_1)\mathbf{k}$$

Hence the area $\triangle OPQ = \frac{1}{2}(x_1y_2 - x_2y_1)$ or $\frac{1}{2}\begin{vmatrix} x_1 & y_1 \\ x_2 & y_2 \end{vmatrix}$

Example (ii)

In a unit cube determine the perpendicular distance from any one corner on to any diagonal which does not pass through it.

Referring to the figure in example (vi) of the section on scalar products, the distance required is the perpendicular distance p of O from $\overrightarrow{BH}$.

Now $|\overrightarrow{BO} \times \overrightarrow{BH}| = OB \sin \angle OBH \times BH$
$= p \times BH$
$= p\sqrt{3}$

But $\overrightarrow{BO} = -\mathbf{i}$ and $\overrightarrow{BH} = \mathbf{j} - \mathbf{i} + \mathbf{k}$

$\therefore \overrightarrow{BO} \times \overrightarrow{BH} = -\mathbf{k} + \mathbf{j}$

and $|-\mathbf{k} + \mathbf{j}| = \sqrt{2}$

Hence $p = \dfrac{\sqrt{2}}{\sqrt{3}}$ or $\frac{1}{3}\sqrt{6}$

Example (iii)

In the triangle OAB, P lies on OA, Q on AB, R on BO (Fig. 79) produced so that $OP:PA = k_1:1$; $AQ:QB = k_2:1$; $BR:OR = k_3:1$. Prove that if P, Q, R are collinear then $k_1k_2k_3 = 1$ (Menelaus's Theorem).

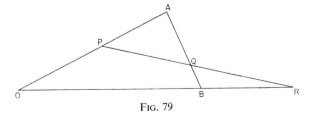

FIG. 79

Represent $\vec{OA}$, $\vec{AB}$, $\vec{BO}$ by vectors $\mathbf{b}$, $\mathbf{c}$, $\mathbf{a}$.

Then $\vec{OP} = \dfrac{k_1}{1+k_1}\mathbf{b}$; $\vec{OQ} = \mathbf{b} + \dfrac{k_2\mathbf{c}}{1+k_2}$; $\vec{OR} = -\dfrac{1}{1-k_3}\mathbf{a}$

and $\vec{PQ} = \vec{OQ} - \vec{OP} = \dfrac{k_2\mathbf{c}}{1+k_2} + \dfrac{1}{1+k_1} \cdot \mathbf{b}$

$\vec{PR} = \vec{OR} - \vec{OP} = \dfrac{-\mathbf{a}}{1-k_3} - \dfrac{\mathbf{b}k_1}{1+k_1}$

P, Q, R are collinear if $\vec{PQ} \times \vec{PR} = 0$.

i.e. if $\left(\dfrac{k_2\mathbf{c}}{1+k_2} + \dfrac{1}{1+k_1}\mathbf{b}\right) \times \left(\dfrac{\mathbf{a}}{1-k_3} + \dfrac{k_1\mathbf{b}}{1+k_1}\right) = 0$

i.e. $\dfrac{k_2\,(\mathbf{c}\times\mathbf{a})}{(1+k_2)(1-k_3)} - \dfrac{k_2 k_1\,(\mathbf{b}\times\mathbf{c})}{(1+k_1)(1+k_2)} - \dfrac{1\,(\mathbf{a}\times\mathbf{b})}{(1+k_1)(1-k_3)} = 0$

$\mathbf{c}\times\mathbf{a} = \mathbf{b}\times\mathbf{c} = \mathbf{a}\times\mathbf{b} =$ twice the vector area of triangle OAB. Hence, dividing throughout by this vector and multiplying by $(1+k_1)(1+k_2)(1-k_3)$ we have

$$k_2(1+k_1) - k_1 k_2(1-k_3) - (1+k_2) = 0$$

i.e. $k_1 k_2 k_3 = 1$

The development of vector theory beyond this point is not the concern of the secondary school, except perhaps for the occasional gifted sixth-form pupil, who may, by private reading, extend his knowledge to cover scalar and vector triple products and the differentiation of vectors. On the elementary aspects of the subject covered so far, however, there is an enormous variety in the number of exercises and applications that may be devised.

One would not suggest that vector methods in geometry be employed to the exclusion of all others, but they should be demonstrated as an alternative system or mathematical tool of real power in suitable situations.

One of the undesirable features of school geometry in the past has been that, in the eyes of pupils at any rate, it has stopped dead at O-level. For the purposes of most single-subject advanced level syllabuses one does not come back into the sixth form and develop the subject any further, except perhaps for Menelaus's and Ceva's theorems or, in an entirely different way, through Cartesian coordinates. Vector methods, however, may be introduced in a simple way at the age of 12 or 13 and developed continuously thereafter right up to degree level if necessary. At every stage new uses and applications arise; it is not a subject that "dries up".

MOTION GEOMETRY

An essentially similar but non-algebraic approach to the teaching of elementary geometry has been developed from the ideas of motion geometry expounded in I. M. Yaglom's now classic work *Geometric Transformations.*

Yaglom gives the following definition of geometry: "Geometry is the science that studies those properties of geometric figures that are not changed by motions of the figures." Further: "A motion is a geometric transformation of the plane (or of space) carrying each point A into a new point A′ such that the distance between any two points A and B is equal to the distance between the points A′ and B′ into which they are carried."

Under the general headings of displacements and symmetries, motion geometry is developed in terms of the four basic motions in the plane, viz. translation, reflection, rotation (half-turn symmetry) and glide reflection. To these may later be added the principle of enlargement. Under each of these motions or operations certain features remain invariant. Under the first four, lengths and angles are unaltered; under enlargement, ratios and angles are invariant. Using these principles it is possible to demonstrate the truths of elementary geometry and some of these demonstrations are astonishingly simple.

Immediate consequences of the principle of symmetry by reflection include the following:

(1) The base angles of an isosceles triangle are equal.
(2) If a line is drawn from the centre of a circle perpendicular to a chord, it bisects the chord.
(3) The angle at the centre of a circle is twice the angle at the

circumference standing on the same arc (together with all the circle theorems which are corollaries of this).

(4) Tangents to a circle from an external point are equal in length.

Using the invariance of lines and angles under the rotation of a geometrical figure about some fixed point in its fixed plane, we quickly obtain such results as:

(5) Properties of parallelograms.
(6) The midpoint theorem.
(7) Equal chords of a circle subtend equal angles at the centre and are equidistant from the centre.

As the name suggests, much of the classroom work in motion geometry may be carried out by the pupils themselves, using chinagraph pencilled diagrams on transparent acetate sheets and moving one sheet upon another. It is an approach by which children can learn and discover a good deal for themselves; intuition is not discouraged and the whole idea of working in terms of basic motions or transformations in the plane fits in beautifully with the ideas we have discussed in the chapters on groups and matrices.

By introducing a principle of enlargement it is possible to obtain the geometrical properties of similar figures. Alternatively, the ratio theorems can all be developed by vectors as in the present chapter.

It must be stressed that the two approaches are by no means mutually exclusive. It may well be that a predominantly vector approach to geometry holds more interest and value for a pupil who is to pursue a career in engineering or technology, while a motion geometry approach may be more suitable for others. Nevertheless, to use one method exclusively would be quite wrong. Children enjoy finding alternative solutions to the same problem, and after all there is no virtue in using only one tool when others are freely available. Our aim in good mathematics teaching should be to teach discrimination in the methods available and an appreciation of elegance and economy in their use.

Useful Reference Books

ALLEN, R. G. D., *Basic Mathematics*, Macmillan.
COXETER, H. S. M., *Introduction to Geometry*, Wiley.
CABLE, J., What is a Vector?, *The Mathematical Gazette*, Feb. 1964, Bell.
DAVIS, H. S., *Introduction to Vector Analysis*, Prentice-Hall.
ELIEZER, C. J., *Concise Vector Analysis*, Pergamon Press.
FLETCHER, T. J., *Some Lessons in Mathematics*, C.U.P.
GILES, G., Vector Geometry in School, *The Mathematical Gazette*, Feb. 1964, Bell.
HAGUE, *Introduction to Vector Analysis*, Methuen.
HARRISON, E., Vectors, *Mathematics Teaching*, No. 22, Spring 1963.
KEMENY, MIRKIL, SNELL and THOMPSON, *Finite Mathematical Structures*, Prentice-Hall.
LOWRY, H. V., and HAYDEN, H. A., *Advanced Mathematics for Technical Students*, Book II, Longmans.
PERFECT, H., *Topics in Geometry*, Pergamon Press.
SMITH, G. D., *Vector Analysis Including the Dynamics of a Rigid Body*, Oxford University Press.
STEWART, C. A., *Advanced Calculus*, Methuen.
WEATHERBURN, C. E., *Elementary Vector Analysis*, Bell.
WEXLER, C., *Analytic Geometry—A Vector Approach*, Addison-Wesley.
WEYL, H., *Symmetry*, Oxford University Press.
WOLSTENHOLME, E. Œ., *Elementary Vectors*, Pergamon Press.
YAGLOM, I. M., trans. by A. Shields, *Geometric Transformations*, Random House New Mathematical Library.

6

INEQUALITIES AND LINEAR PROGRAMMING

LATTICES

In Chapter 1 we discussed the representation of sets of ordered pairs and the manner in which a unit square lattice of points may arise as the solution set of an inequality of the form $ax+by > c$, where x and y are elements of the set of integers.

In the present chapter we develop the idea further and show that this type of work can lead to a class of most interesting and practical problems. We shall use the matrix notation where appropriate and show that one of the by-products of the development is a need to solve simultaneous linear equations. In the classroom this need may be exploited in order either to introduce the inverse matrix, or, perhaps, to provide an opportunity for useful revision. The more we can relate topics in mathematics, the more coherent and meaningful does the subject become. The modern approach through sets, groups and matrices is particularly susceptible to such interrelationship.

Consider the set of ordered pairs $\{(x,y) | x+y \geq 4\}$, where x and y are elements of J, the set of integers. This set may be plotted graphically or on a pegboard and we find the result appears as shown in Fig. 80.

The set is, of course, infinite. Furthermore it consists of the union of two sets; (i) the ordered pairs $\{(x,y) | x+y = 4\}$, and (ii) the ordered pairs $\{(x,y) | x+y > 4\}$. The first set consists of the ringed dots, the second of the unringed dots: the whole set is

$$\{(x,y) | x+y = 4\} \cup \{(x,y) | x+y > 4\}$$

INEQUALITIES AND LINEAR PROGRAMMING 133

We now investigate what happens when x and y may be integers or halves of integers. Obviously the number of dots is quadrupled and the "mesh" of the lattice is halved.

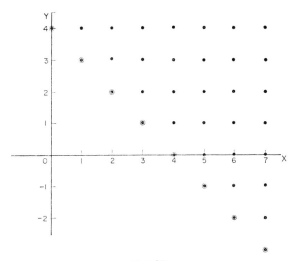

FIG. 80

Suppose now that x and y may take any rational values. Immediately the pattern of dots becomes very dense indeed. Does it become so dense that the space between the original dots is completely filled? After all, given any two rational numbers we can always find a third one which lies between them. For suppose that $\dfrac{p}{q} < \dfrac{r}{s}$, where p, q, r, s are positive integers,

then
$$ps < qr$$
$$\therefore ps + pq < qr + pq$$
$$\therefore p(q+s) < q(p+r)$$
$$\therefore \frac{p}{q} < \frac{p+r}{q+s} \tag{1}$$

also
$$ps+rs < qr+rs$$
$$\therefore s(p+r) < r(q+s)$$
$$\therefore \frac{p+r}{q+s} < \frac{r}{s} \tag{2}$$

(1) and (2) give $\dfrac{p}{q} < \dfrac{p+r}{q+s} < \dfrac{r}{s}$

i.e. no matter how small the difference between $\dfrac{r}{s}$ and $\dfrac{p}{q}$ we can always find a rational fraction $\dfrac{p+r}{q+s}$ which lies between them. For example between $\frac{1}{2}$ and $\frac{1}{3}$ there lies $\frac{2}{5}$, between $\frac{1}{2}$ and $\frac{2}{5}$ there lies $\frac{3}{7}$, between $\frac{1}{2}$ and $\frac{3}{7}$ there lies $\frac{4}{9}$, etc.

This is an ideal point at which to introduce the concept of irrational numbers. The assumption that $\sqrt{2}$ can be expressed as a rational fraction p/q in its lowest terms implies that $p^2 = 2q^2$. $2q^2$ is even, hence p^2 is even and hence p is even. Write $p = 2k$ where k is an integer. We now have $4k^2 = 2q^2$, therefore q^2 and q are even. However, p and q cannot both be even if p/q is in its lowest terms. The original assumption leads to a contradiction and hence there exist numbers which are not rational fractions. These numbers are called irrational numbers. If we now include these amongst the permissible values of x and y we do, in fact, fill the plane on and above the line $x+y = 4$, and instead of a lattice we have a continuum of points or ordered pairs which is a subset of the product set $R \times R$ where R is the set of real numbers.

A discussion of this type will obviously do more harm than good with *young* children, and if linear programming problems are introduced early in the secondary school syllabus it is better to assume at the outset that real number pairs fill the plane, rather than attempt to discriminate between rational and irrational numbers.

OPEN AND CLOSED HALF-PLANES

For the set of real numbers therefore, the set $\{(x,y) \mid x+y \geqq 4\}$ is called the half-plane. This set includes the set of points on the line $x+y = 4$, i.e. it includes its "boundary". The set of points for which $x+y > 4$, does not include the boundary. In the first

INEQUALITIES AND LINEAR PROGRAMMING 135

case we say that the solution set is the *closed half-plane*; when $x+y > 4$ we have the *open half-plane*. This corresponds to the idea of a *closed* interval such as $0 \leq x \leq 2$ and an *open* interval such as $-1 < x < 1$.

The sets $\{(x,y)|x+y > 4\}$, $\{(x,y)|x+y = 4\}$, $\{(x,y)|x+y < 4\}$ are disjoint subsets of the product set $R.R$ where R is the set of real numbers; their union is, of course, the entire plane.

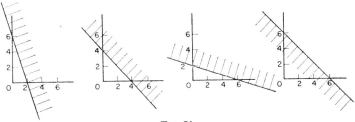

Fig. 81

The next step is to consider the intersection of two or more sets of this form. Suppose we have the closed half-planes corresponding to the solution sets of the following inequalities:

$$3x+y \geq 6, \quad x+y \geq 4, \quad x+3y \geq 6, \quad x+y \leq 6$$

In what region are these satisfied? The respective half-planes are shown shaded in the diagrams of Fig. 81 and their intersection appears as the shaded polygon of Fig. 82.

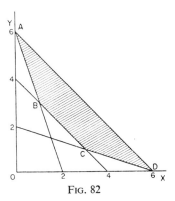

Fig. 82

CONVEX SETS

For two variables this intersection is called a *polygonal convex set*. For one variable it would simply become a closed interval, but for three variables we should have half *spaces* intersecting in a polyhedral space. Such a space is called a *polyhedral convex space*. Some writers use the term polyhedral convex set for cases of all dimensions, but since we shall restrict our examples to two-dimensional problems we shall refer simply to such intersections as *convex sets*.

In the previous diagram there are four points A, B, C, D. These are called *extreme points* of the convex set. It is a simple matter to determine them. B, for example, is the intersection of the sets $\{(x,y) | 3x+y = 6\}$ and $\{(x,y) | x+y = 4\}$. A more familiar way of putting this is to say that the coordinates of B are the solutions of the simultaneous equations

$$3x+y = 6$$
$$x+y = 4$$

or
$$\binom{x}{y} = \frac{1}{2}\begin{pmatrix} 1 & -1 \\ -1 & 3 \end{pmatrix} \cdot \binom{6}{4}$$
$$= \frac{1}{2}\binom{2}{6} = \binom{1}{3}, \quad \text{i.e. } x = 1, \quad y = 3$$

although in this particular case the solution is obvious.

We shall see later that if it is further required to find the maximum or minimum values of some function $ax+by$ subject to the conditions already given, then such maximum or minimum values occur at these extreme points.

LINEAR PROGRAMMING

In the highly competitive markets of the modern world, industrial and commercial enterprises are obliged to ensure that their organizations are such that certain factors such as productivity and profit are as large as possible, while wastage, transport and manufacturing costs generally are kept as low as possible. The individual conditions required to achieve this state of affairs are frequently simple and linear in form. There are usually so many variables though that the number of conditions or linear type inequalities is very great. Maximizing the profit subject to a hundred linear conditions concerning shipping costs, transport costs, packaging costs, fuel costs, orders taken, sales expected, overtime rates, etc., is a job for the computer and not for the schoolboy. On the other hand, it is a situation which he readily appreciates. His own mother sometimes has to decide whether to shop locally, saving time and bus fares, or whether to go into town where the market prices are lower and the goods seem fresher.

The following examples are greatly simplified. On the other hand, they are still meaningful; they are real-life problems. Children seem to find them worth doing and in doing them they learn and revise many techniques. They also have to think intelligently. In this type of investigation there is more than one "right" answer, and occasionally more data than necessary is given. Intelligence and discrimination are required in addition to mere technique.

Example 1

A master printer employs journeymen and apprentices and his facilities are such that he cannot employ more than 9 people altogether. His orders oblige him to maintain an output of at least 30 units of printing work per day. On the average a journeyman does 5 units of printing work and an apprentice 3 units of work daily. The Apprentices Act demands that the printer should employ not more than 5 men to 1 apprentice. The journeyman's union, however, forbids him to employ less than 2 men to each apprentice. How many journeymen and apprentices should he employ? Of the possible results, which is the wisest choice if he has to pay journeymen £2 per day and apprentices £1 per day? If he charges £1 per printing unit to his customers and can sell all his output in excess of 30 units, in which case is his gross profit a maximum?

Let us suppose that he employs x journeymen and y apprentices. The conditions to be met reduce to the following inequalities:

$$5x+3y \geq 30 \text{ (output at least 30 units per day)} \qquad (l_1)$$
$$x+y \leq 9 \quad \text{(not more than 9 employees altogether)} \quad (l_2)$$
$$y \geq \tfrac{1}{5}x \quad \text{(Apprentices Act)} \qquad (l_3)$$
$$y \leq \tfrac{1}{2}x \quad \text{(union rules)} \qquad (l_4)$$

or in matrix form

$$\begin{pmatrix} 5 & 3 \\ -1 & -1 \\ -1 & 5 \\ 1 & -2 \end{pmatrix} \cdot \begin{pmatrix} x \\ y \end{pmatrix} \geq \begin{pmatrix} 30 \\ -9 \\ 0 \\ 0 \end{pmatrix}$$

The intersection of the solution sets of these inequalities is shaded in Fig. 83.

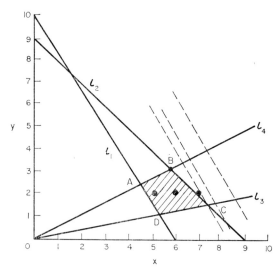

Fig. 83

INEQUALITIES AND LINEAR PROGRAMMING

Any ordered pair (x,y) on either the interior or the perimeter of the convex set, vertices A, B, C, D, satisfies the required conditions. In this case, however, x and y may only take integral values, and of these there are two pairs on the interior (5 men, 2 apprentices), (6 men, 2 apprentices); one on BC (7 men, 2 apprentices); and one at the extreme point B (6 men, 3 apprentices).

In order to decide which is the wisest we might analyse these possibilities as in Table 29.

TABLE 29

Men (x)	Apprentices (y)	Total daily wages £	Total daily output (units)	Daily takings	Output per £1 wages	Gross daily profit £
5	2	12	31	31	2·58	19
6	2	14	36	36	2·57	22
6	3	15	39	39	2·60	24
7	2	16	41	41	2·56	25

By employing 6 men and 3 apprentices there is a slightly higher output per £1 wages paid, but by charging £1 per unit for the finished work, the employer's gross daily profit is greatest for 7 men and 2 apprentices. Under the circumstances, therefore, he will probably employ 7 men and 2 boys.

It is interesting to note that if he could only sell his output at 10s. per unit, his gross daily profit in each of the last two cases would be £4. 10s. and there would be nothing to choose between the pairs (6,3) and (7,2). Further, if he could only sell at 9s. per unit, his gross daily profits corresponding to the cases (6,3) and (7,2) would be £2. 11s. and £2. 9s. respectively. If he is forced to charge less than 10s. per unit it would therefore be better to employ 6 men and 3 boys.

Let us forget for a moment that x and y in this particular case must take only integral values. There are now an infinite number of solutions satisfying the original inequalities. Suppose, however, that we wish to obtain one for which the gross daily takings £T is a maximum. Selling at £1 per unit we have $5x+3y = T$. How can we maximize the function $5x+3y$ or T over the convex set satisfying the inequalities?

Referring once again to the diagram, we see that $5x+3y = T$ is a family of parallel straight lines for varying values of T. Three members of this family are shown as dotted lines. The further one of these lines is from the origin, the greater is the value of T. If, however, the inequalities are to be satisfied, the line must contain a member of the convex set. In order to do this for the maximum value of T we must select that line which passes through C.

It is easy to check that C is the point $(7\frac{1}{2}, 1\frac{1}{2})$. In this case the gross takings would be £$(5 \times 7\frac{1}{2} + 3 \times 1\frac{1}{2})$ or £42. In other words, the theoretical maximum of the function $5x+3y$ taken over the convex set A, B, C, D occurs at one of its extreme points C.

If it were possible to employ $7\frac{1}{2}$ men and $1\frac{1}{2}$ boys, the gross profit in this case would also be larger, viz. £25. 10s.

Of course, were we to double the size of the whole firm, allowing a maximum of 18 employees instead of only 9, then the theoretical maximizing solution of $7\frac{1}{2} : 1\frac{1}{2}$ would also be the practical one of $15:3$, or 15 men to 3 apprentices.

In simple cases such as this we usually find that the maximum occurs at one extreme point only of the convex set. Referring to the diagram again; imagine that instead of trying to maximize the function $5x+3y$ over this set, we were trying, for some reason or other, to maximize the function $x+y$. Drawing parallel lines $x+y = c$, we should find that immediately the line $x+y = c$ included one point of the convex set, it would include a *whole line segment* of $x+y = 9$.

Thus maximizing and minimizing solutions usually occur at extreme points of a convex set, but in certain cases it is possible to get a whole line segment of maximizing solutions. Such a case arises in Example 4.

The reader may care to attempt the following examples without reference to the solutions. These are provided in full for those who prefer to read on, or alternatively, as a check on the reader's own solutions.

Example 2

A dictator seizes power in a small state and proceeds to plan the economy and labour forces. He discovers that there are two motor corporations. Each factory owned by corporation A manufactures weekly 30 vans, 10 saloon cars and 10 lorries, while each factory of corporation B makes 10 vans, 10 saloon cars and 40 lorries weekly. He finds that the average combined home and overseas market for these vehicles is at least 100 vans, 60 saloon cars and 120 lorries weekly, but that previously these demands have been greatly exceeded in some respects and not met in others. In his reorganization, how many factories in each corporation should continue to operate? If the labour force in each factory of corporation A is half that in each factory of corporation B, and if the profits are roughly proportional to the labour force, which is the better decision if (a) he wishes to economize on labour, (b) he wishes to maximize the profits.

Suppose that there are x factories in corporation A and y factories in corporation B. We may set out the conditions as follows:

	Corporation A	Corporation B
Vans	30	10
Saloons	10	10
Lorries	10	40

$$\begin{pmatrix} 30 & 10 \\ 10 & 10 \\ 10 & 40 \end{pmatrix} \cdot \begin{pmatrix} x \\ y \end{pmatrix} \geq \begin{pmatrix} 100 \\ 60 \\ 120 \end{pmatrix}$$

or separately

$$3x + y \geq 10 \quad (l_1)$$

$$x + y \geq 6 \quad (l_2)$$

$$x + 4y \geq 12 \quad (l_3)$$

and the intersection of the solution sets is shown as a shaded area in Fig. 84.

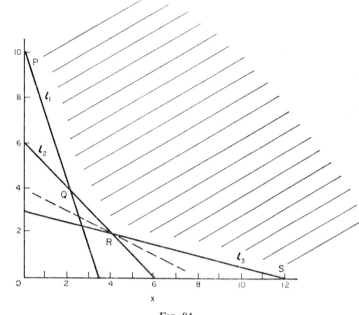

Fig. 84

The extreme points are P, Q, R, S.

At P, $x = 0$, $y = 10$
At Q, $x = 2$, $y = 4$
At R, $x = 4$, $y = 2$
At S, $x = 12$, $y = 0$

and all of them satisfy the inequalities.

Assume now that corporation A employs 500 workers in each factory and corporation B 1000 workers, and that the profits per factory are respectively 500 and 1000 units. We now have the results shown in Table 30.

TABLE 30

	Total workers employed	Total profits
At P	10,000	10,000
Q	5,000	5,000
R	4,000	4,000
S	6,000	6,000

At P and S the estimated market is greatly exceeded. For example, at P we should have 400 lorries, and at S 360 vans! A wise choice, therefore, is either (a) R, from the point of view of economizing on the labour force committed to the industry, or (b) Q from the point of view of maximum realizable profits.

Minimizing the labour force of course means minimizing the function $500x+1000y$, or the function $x+2y$ over the convex set. In order to do this we seek the line $x+2y = L$ which is *nearest* to the origin and yet contains at least one point of the convex set. The line required is shown dotted in the diagram and it is clear that the extreme point R is the minimizing one for economizing on the labour force.

Example 3

A dietitian wishes to mix together two kinds of food so that the vitamin content of the mixture is at least 9 units of vitamin A, 7 units of vitamin B, 10 units of vitamin C and 12 units of vitamin D. The vitamin content per pound of each food is shown below:

	Vitamin A	Vitamin B	Vitamin C	Vitamin D
Food 1	2	1	1	1
Food 2	1	1	2	3

If food 1 costs 5s. per lb and food 2 costs 7s. per lb, find the minimum cost of such a mixture.

When x lb of food 1 is mixed with y lb of food 2, the conditions regarding vitamin content reduce to the matrix inequality:

$$\begin{array}{c} \text{Vitamin A} \\ \text{Vitamin B} \\ \text{Vitamin C} \\ \text{Vitamin D} \end{array} \begin{pmatrix} \text{Food 1} & \text{Food 2} \\ 2 & 1 \\ 1 & 1 \\ 1 & 2 \\ 1 & 3 \end{pmatrix} \cdot \begin{pmatrix} x \\ y \end{pmatrix} \geq \begin{pmatrix} 9 \\ 7 \\ 10 \\ 12 \end{pmatrix}$$

or

$$2x+y \geq 9 \qquad (l_1)$$
$$x+y \geq 7 \qquad (l_2)$$
$$x+2y \geq 10 \qquad (l_3)$$
$$x+3y \geq 12 \qquad (l_4)$$

The function to be minimized is $5x+7y$.

In Fig. 85 the intersection of the various solution sets, or the convex set, is shown as a shaded area and two members of the

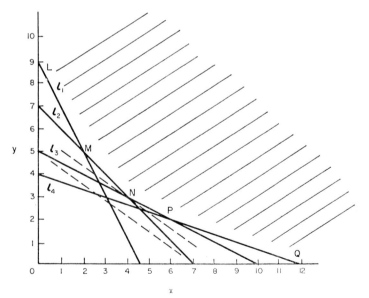

Fig. 85

INEQUALITIES AND LINEAR PROGRAMMING

family of straight lines $5x + 7y = C$ are shown as dotted lines. The member of this family which has the least value of C, while containing at least one point of the convex set, is that line which passes through the extreme point N.

At N, $x = 4$ and $y = 3$. Four lb of food 1 mixed with 3 lb of food 2 satisfies the required vitamin content. It is also the smallest weight which does so, and, of course, it costs the least. The results corresponding to the extreme points L, M, N, P, Q are shown in Table 31.

TABLE 31

	Food 1	Food 2	Weight of mixture	Cost of mixture (shillings)
L	0	9	9	63
M	2	5	7	45
N	4	3	7	*41*
P	6	2	8	44
Q	12	0	12	60

$41s.$ is the minimum cost of making the required mixture.

Example 4

A mine manager has contracts to supply weekly: 1000 tons of grade 1 coal; 700 tons grade 2; 2000 tons grade 3 and 4500 tons of grade 4 coal. Seam A and seam B are being worked at a cost of £4000, £10,000 respectively per shift, and the yield in tons per shift from each seam is given by the table below:

	Grade 1	Grade 2	Grade 3	Grade 4
Seam A	200	100	200	400
Seam B	100	100	500	1500

How many shifts per week should each seam be worked in order to fill the contracts most economically?

Suppose that seam A is worked x shifts per week and seam B y shifts per week. We then have the matrix inequality:

$$\begin{array}{c} \text{Grade 1} \\ \text{Grade 2} \\ \text{Grade 3} \\ \text{Grade 4} \end{array} \begin{pmatrix} \overset{\text{Seam}}{\underset{A}{}} & \overset{\text{Seam}}{\underset{B}{}} \\ 200 & 100 \\ 100 & 100 \\ 200 & 500 \\ 400 & 1500 \end{pmatrix} \cdot \begin{pmatrix} x \\ y \end{pmatrix} \geq \begin{pmatrix} 1000 \\ 700 \\ 2000 \\ 4500 \end{pmatrix}$$

or the set of inequalities:

$$2x+y \geq 10 \qquad (l_1)$$
$$x+y \geq 7 \qquad (l_2)$$
$$2x+5y \geq 20 \qquad (l_3)$$
$$4x+15y \geq 45 \qquad (l_4)$$

The running costs per week of the whole mine amount to £$(4000x+10,000y)$. It is required, therefore, to minimize the function $2x+5y$ over the convex set satisfied by the matrix inequality.

Extreme points of the convex set (Fig. 86) are D, E, F, G, H. The parallel dotted lines are all members of the family of straight lines $2x+5y = C$. When we come to look for that member which is nearest to the origin (minimum C), but which satisfies the inequalities, we find that it coincides with the whole segment FG. We thus have a line segment of minimizing solutions to the problem.

The running costs corresponding to the extreme points D, E, F, G, H are shown in Table 32.

TABLE 32

	Seam A (x)	Seam B (y)	Weekly running cost £
D	0	10	100,000
E	3	4	52,000
F	5	2	40,000
G	7½	1	40,000
H	11¼	0	45,000

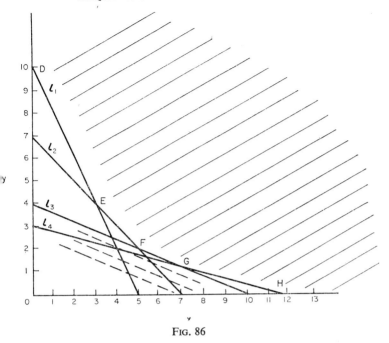

Fig. 86

As we expect, F and G both give minimizing pairs of values. Further, for any point on FG, say $(x_1\ y_1)$, $2x_1 + 5y_1 = 20$. Therefore £$(4000x_1 + 10{,}000y_1) = £40{,}000$.

In practice, of course, the manager will not be solely concerned with the operating costs; the quality of the output, sale price and actual profit will be of more interest to him than operating costs alone. Without making detailed calculations for various prices of the various grades of coal we may compare the nature of the output in tons for F and for G:

	Grade 1	Grade 2	Grade 3	Grade 4
F (5, 2)	1200	700	2000	5000
G (7½, 1)	1600	850	2000	4500

Total output at F is 8900 tons and at G is 8950 tons. Furthermore, the output corresponding to G contains a higher proportion of better grade coal. It is likely, therefore, that in order to achieve maximum profits and minimum operating costs, the manager will decide to operate seams A and B for $7\frac{1}{2}$ shifts and 1 shift per week respectively.

THE TRANSPORT PROBLEM

In most industrial organizations transport costs have to be taken into account and minimized where possible, but in some industries the expense of distribution may well be a major consideration. To oil companies, for example, the economic operation of tanker fleets on land and sea is obviously vital. In order to achieve this they base their schedules upon the results of linear programming.

In the following example we discuss two methods of solving an extremely simple transport problem.

Example 5

A distributor of certain manufactured articles (units) supplies orders from two depots D_1 and D_2. At a certain time D_1 and D_2 have in stock 80 and 20 of these articles respectively. Two customers C_1 and C_2 place orders for 50 and 30 units respectively. The transport cost of any article is directly proportional to the distance it is conveyed, and the distance in miles between the depots and the customers is given in the following table:

	C_1	C_2
D_1	40	30
D_2	10	20

From which depots should the orders be dispatched in order to minimize transport costs?

Let us suppose that the dispatch clerk receives C_2's order for 30 units first. D_2 is nearer to C_2 than D_1 and so he may well decide to send the 20 units from D_2 and the other 10 from D_1. When C_1's order is received the 50 articles must be sent from D_1. His decision may be summarized in matrix form, the transport costs per article being shown as ringed numbers in Table 33.

TABLE 33

	C_1	C_2	
D_1	50 ㊵	10 ㉚	80 in stock
D_2	↓ ⑩	↑ ⑳ 20	20 in stock
	50 ordered	30 ordered	

The total transport cost in this case will be

$50 \times 40 + 10 \times 30 + 20 \times 20 = 2{,}700$ units.

On examining Table 33 carefully we note that if one article is transferred from square $C_1 D_1$ to square $C_1 D_2$, then one article must be transferred from square $C_2 D_2$ to square $C_2 D_1$. (D_2 only contains 20 articles.) Due to the first transfer 30 units of cost will be saved, and due to the second 10 units will be lost. By making this change we save 20 units. The table, therefore, does not represent the most economic solution; each double transfer made in the direction of the arrows reduces the total cost by 20 units of cost per article. The maximum double transfer is 20 articles and if this is done we should have the most economic delivery schedule:

	C_1	C_2
D_1	30 ㊵	30 ㉚
D_2	20 ⑩	0 ⑳

The total transport cost for this arrangement would be

$$30 \times 40 + 30 \times 30 + 20 \times 10 = 2300 \text{ units},$$

which, of course, represents a saving of 20×20 units.

This method is applicable to more complex situations. Given n depots and m customers and their respective stocks and orders, we form a matrix as shown in Table 34.

TABLE 34

	C_1	C_2	C_3	$\ldots C_m$
D_1				
D_2				
D_3				
$\vdots$ D_n				

Starting with any possible delivery schedule, we commence at the top left-hand corner to minimize the cost over four squares as we have done above. We proceed systematically inspecting groups of four squares until the whole matrix has been transformed into one giving the most economic solution.

In example 5 we may also obtain a solution by the graphical method employed in previous examples.

Assume that x articles are delivered from D_1 to C_1 and y articles from D_1 to C_2. The matrix then appears as in Table 35.

TABLE 35

	C_1	C_2	
D_1	x ㊵	y ㉚	80 in stock
D_2	$50 - x$ ⑩	$30 - y$ ⑳	20 in stock
	50 ordered	30 ordered	

The values of x and y are limited by the stocks of D_1 and D_2:

$$x+y \leqq 80 \qquad (1) \quad (l_1)$$
$$(50-x)+(30-y) \leqq 20 \qquad (2)$$
or
$$x+y \geqq 60 \qquad (2) \quad (l_2)$$

Further, in view of the orders placed by C_1 and C_2:

$$0 \leqq x \leqq 50 \qquad (3) \quad (l_3)$$
$$0 \leqq y \leqq 30 \qquad (4) \quad (l_4)$$

The cost of transport is

$$40x+30y+10(50-x)+20(30-y)$$
or
$$(1{,}100+30x+10y) \text{ units}$$

In other words we require to minimize the function $3x+y$ subject to the conditions 1–4.

From Fig. 87 we see that a minimizing solution is given at the point A where $x = y = 30$.

In this problem we note that inequality (1) is superfluous and does not affect the solution; it is the sum of (3) and (4) and

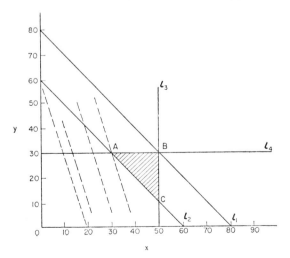

Fig. 87

therefore is not an independent condition. Situations do arise in which even an independent condition does not affect the solution. It is sometimes a good plan to include such cases in exercises. Although it seems to be a convention that in every public examination question the data should be just sufficient for a solution, it is not a practice which helps pupils to discriminate between relevant and irrelevant facts and data.

The solution $x = y = 30$ in this problem gives us the most economic delivery schedules as

	C_1	C_2
D_1	30	30
D_2	20	0

which accords with the results obtained by the alternative method.

Problems of the type we have discussed can be devised by the teacher and they do arouse interest and provide scope for the exercise and revision of many skills. The solution of simultaneous linear equations, the handling of inequalities, the use of matrix notation, the drawing of a straight line, either from two points known to be upon it, or, at a later stage, by inspection of its gradient and constant term, are all contained in such exercises.

The whole subject might, of course, be used as an introduction to the idea of sets, their unions and intersections, instead of the reverse approach used in this chapter.

For example, if $n(P)$ denotes the number of elements (say ordered pairs) in set P, then the formula

$$n(P \cup Q) = n(P) + n(Q) - n(P \cap Q)$$

can be established from the type of diagram we have been considering by looking at the shaded areas in the diagrams of Fig. 88.

We have limited our work to linear programming in two dimensions. Although with three variables it is possible to construct a model showing the convex polyhedron formed by the intersection of half spaces it is hardly practicable; with four or more variables the problem becomes purely algebraic. In actual

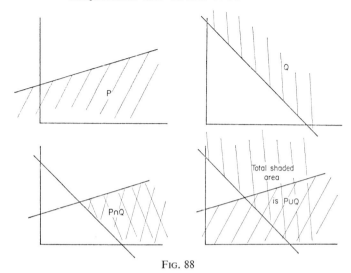

Fig. 88

practice the number of variables may be very large indeed and so a graphical solution is hardly ever possible in a real life problem. Basically however, the algebraic situation is unchanged. Instead of solving simultaneous equations in two unknowns, the real life problem requires solution, usually by computer, of n simultaneous equations in n unknowns where n is large.

NON-LINEAR PROGRAMMING

As an extension of this type of graphical work and to provide practice in the handling of non-linear inequalities and the solution of non-linear equations, problems of the following type might be attempted.

Example 6

Find integral values of x and y satisfying the inequalities

$$xy > 8$$
$$4y - 3x > 0$$
$$x^2 + y^2 < 25$$
$$y^2 < 4x$$

A purely algebraic solution of these inequalities would be beyond the capabilities of O-level candidates, but a graphical solution is quite simple and provides an opportunity of introducing or revising graphs of the circle, parabola and hyperbola and the solution of non-linear simultaneous equations.

In Fig. 89 the intersection of the solution sets (for positive real values of x and y) of the inequalities above is indicated by the shaded area.

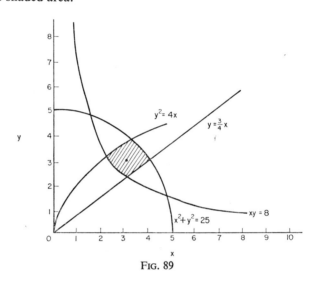

Fig. 89

There is clearly an infinity of real number pairs satisfying the inequalities but only one pair of positive integral values (3, 3).

Problems of this type arise in mensuration. For example, if a drawing-board is to be constructed in such a way that its surface area is greater than 8 ft², its diagonal less than 5 ft and the ratio of its sides not less than 3:4, we should have for the dimensions x', y' the first three of the four inequalities given above.

Linear programming is developed far beyond the type of problem discussed in this chapter. It can be shown, for example, that to every maximizing problem there exists a dual problem which is a minimizing one. To those who wish to pursue the subject beyond this point, the duality theorem and the simplex method are fully discussed in the book *Finite Mathematical Structures* mentioned below.

Useful Reference Books

BRISTOL, J. D., *An Introduction to Linear Programming*, Harrap.
CRANK, J., *Mathematics and Industry*, Oxford University Press.
FLETCHER, T. J., *Some Lessons in Mathematics*, C.U.P.
GALE, D., *The Theory of Linear Economic Models*, McGraw-Hill.
HADLEY, G., *Linear Algebra*, Addison-Wesley.
KEMENY, MIRKIL, SNELL and THOMPSON, *Finite Mathematical Structures*, Prentice-Hall.
MANSFIELD, D. E., and THOMPSON, D., *Mathematics, a New Approach*, Book II, Chatto & Windus.
REYNOLDS, J. A. C., *Shape, Size and Place*, E. Arnold.
RICHARDSON, *Fundamentals of Mathematics*, Macmillan.
VAJDA, S., *Introduction to Linear Programming and the Theory of Games* Methuen.

7
STATISTICS

It is hardly necessary to explain what statistics are. One has only to open the paper or watch television to see figures relating to unemployment, party popularity polls or the latest "top ten" discs. The amassing of statistical information has been the concern of governments for centuries. At the time of Christ's birth a general census was being undertaken; in this country today the Central Statistical Office is exclusively concerned with the regular collection, tabulation, analysis and publication of a huge range of meteorological, geographical, industrial and vital statistics.

Apart from this, nearly every manufacturing organization of any size employs its own statistician or statistical department. It is clearly impossible for a manufacturer of nylon yarn to test every inch of the product he sells, but by selective sampling of random batches by methods based on sound mathematical analysis, the statistician is able to ensure that certain tolerances regarding, say, tensile strength are not exceeded. Any adverse trend will soon become apparent and the manufacturing fault will be traced before too much time and raw material have been wasted.

Medical units are constantly carrying out checks on the incidence of diseases such as bronchitis and its correlation with air pollution and other factors likely to affect it.

Unless analysed properly statistics can be meaningless or even positively misleading. Political views and advertisers' claims for their products are sometimes supported by specious statistical arguments. In the past this type of misuse has occurred to such an extent as to create a certain mistrust of the subject. "There are lies, damned lies and statistics."

It is surely a proper part of every child's education to learn

STATISTICS 157

something of the way in which large sets of data may be categorized and sifted, and to learn how to discriminate between true and false conclusions that might be drawn from them.

To a degree this is already being done. Children collect data such as rainfall and temperature changes and learn to represent it graphically. Some examination boards have introduced a statistical section into additional mathematics papers at O-level, and many boards offer statistics as an alternative to mechanics at A-level. In the writer's view this is not a wholly satisfactory state of affairs. It is still possible for a child to embark upon a post-A-level course in economics or biology without having encountered the first principles of statistics; in any other branch of science or mathematics it is more likely than ever that this situation will obtain.

When we come to look at it, however, elementary statistics is a highly suitable topic for inclusion in O-level syllabuses. Apart from general educational considerations it provides another way of looking at the elements of a set. The techniques required to calculate, for example, a standard deviation, include addition, multiplication, division and the extraction of the square root. This may be taken as useful revision of previous work or as a first-class opportunity of introducing the use of simple calculating machines. In order to start the work some initial sorting of data may be involved, and when the analysis has been completed it may be possible to arrive at certain conclusions. The net result is often interest and enthusiasm and the knowledge acquired is useful to all and invaluable to some.

In this chapter we give an elementary introduction to the subject as it might be taught in secondary schools.

COLLECTION OF DATA

Data must be collected with a clearly defined aim in view. If, for example, we were conducting an inquiry into household budgets, it would be necessary to realize exactly what we were trying to discover. A more limited objective is easier to achieve than a complicated one. Suppose we wish to find what proportion of the average family income goes on food, rent or mortgage repayments, clothing, drink and tobacco, entertainment, books and periodicals. Clearly we cannot canvass the whole population; on the other hand our sample must be (a) large enough, (b) sufficiently representative. If we decide to elicit the information by means of a questionnaire, the questions must be simply and unambiguously worded. Further, they should be framed in such a way that they may be answered "Yes" or "No"; or by a numerical quantity. Finally, we should realize, and be prepared to allow for, the risk of false information.

A survey was once conducted to find the ideal length of a questionnaire. It was found that normally a questionnaire should occupy more than one page but preferably less than two; it should be necessary to turn over. A questionnaire which is too short is hardly worth filling in at all. On the other hand, there is a limit to the time that one is prepared to spend on such things; or so it seemed.

It is a *sine qua non* that an inquiry should seem to be serving a worthwhile purpose. Questions should be courteously phrased and should avoid asking for information which could place the person completing the questionnaire in a compromising situation.

A good deal of statistical analysis is, of course, carried out on information which is obtained by experiment or observation. The scatter of results obtained in a physical experiment by a highly trained scientist is a case in which the data will be extremely reliable, but the "answer" cannot possibly be "true". The best that can be achieved is a good approximation to the unknown result. On the other hand, a traffic survey of vehicles using the M2 on a particular day is a case in which there is a "true" answer. In this case the data will be as reliable as the observers. This is likely to be of a fairly high order and corresponding reliance can be placed on the results.

REPRESENTATION OF DATA

Information, when collected and categorized, may be presented as tabulated figures. For popular consumption and greater impact, however, results are sometimes reduced to pictorial or diagrammatic form.

STATISTICS

As an example, suppose that during a certain year the national expenditure on domestic budgets was as shown in Table 36.

TABLE 36

	(£m)†
Food	4500
Housing	1200
Clothing	1200
Tobacco and drink	1700
Entertainment and books	400

† These figures are roughly proportional to the actual figures for 1958.

These may be represented diagrammatically in various ways.

(a) *The pie-chart or circular diagram* (Fig. 90). The figures given are proportional to the numbers 45, 12, 12, 17, 4—total 90. Representing one part by 4 degrees of angle we divide a circle into 5 sectors of angle 180°, 48°, 48°, 68°, 16°. Labelling the appropriate sectors we have the diagram shown in Fig. 90.

FIG. 90

(b) *The bar-chart* (Fig. 91). Alternatively we may represent the expenditure by columns of equal width with their heights proportional to the amounts spent. If the food column is 4·5 in. high, the other columns must be 1·2 in., 1·2 in., 1·7 in. and 0·4 in. high respectively.

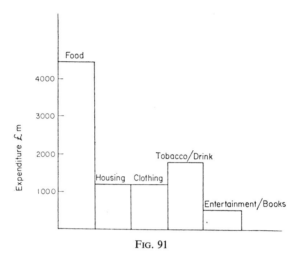

Fig. 91

(c) *Pictorial representation* (Fig. 92). Choosing pictorial units which correspond in each case to, say, £400 million expenditure, we might have Fig. 92.

Fig. 92

FREQUENCY POLYGONS AND HISTOGRAMS

In both methods we represent the independent variable quantity investigated on a horizontal axis, and the frequency with which particular values occur along a vertical axis. Where discrete values only of the variable occur we sometimes use a frequency polygon.

Suppose, for example, that twelve people were asked to guess the weight of a cake to the nearest half-pound, and that the answers obtained were as follows:

$3\frac{1}{2}$, 3, 4, $4\frac{1}{2}$, $3\frac{1}{2}$, 3, 5, 4, $3\frac{1}{2}$, 2, $3\frac{1}{2}$, 4   lb.

Classifying the results we have

Guess (lb)	Frequency
(x)	(f)
2	1
3	2
$3\frac{1}{2}$	4
4	3
$4\frac{1}{2}$	1
5	1

Represented as a frequency polygon they appear as Fig. 93.

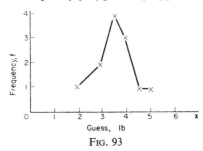

Fig. 93

By this means we see at a glance the way in which the results are distributed, and it is reasonable to expect that the actual weight lies somewhere between 3 and 4 lb.

The most popular guess is $3\frac{1}{2}$ lb. We call this the *mode* or *modal value*. The *average* or *mean* guess is

$$\frac{2 \times 1 + 3 \times 2 + 3\frac{1}{2} \times 4 + 4 \times 3 + 4\frac{1}{2} \times 1 + 5 \times 1}{12} \quad \text{or} \quad 3 \cdot 625 \text{ lb.}$$

Sometimes we refer to the *median* value. This is the middle value of the variables arranged in order of magnitude. For twelve guesses the median value is the one corresponding to the 6th or 7th guess, and in this case both of these are $3\frac{1}{2}$ lb. Generally the mode, mean and median values are different, but in perfectly symmetrical distribution they do, of course, coincide.

The frequency polygon is of limited value. Consider, for example, an inquiry into the annual income of forty men. It is unlikely that any two receive exactly the same salary. Each individual salary would probably appear only once in the analysis, and in such a case the frequency polygon would consist of a horizontal straight line through frequency $(f) = 1$. Any pattern that did exist would certainly not be revealed by this means.

Suppose, however, we form the men into certain income groups, and represent the frequency of each group (i.e. the number of men in each group) by a column as in a bar-chart. We should then have a *histogram*. Thus, if the income groupings were as follows:

£0–499	8 men
£500–999	15 men
£1000–1499	11 men
£1500–1999	4 men
£2000–2499	2 men

the histogram would appear as Fig. 94.

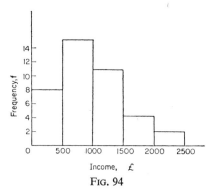

FIG. 94

Forty men is a relatively small sample and we should not expect to discover any pattern relating to incomes in general. It might be that these men were members of a club whose membership is not representative of the general community. Nevertheless, we get an impression of heavy grouping around the £750 mark and a rapid tailing off as we reach the higher income bracket.

In a case like this, where we do not know the individual incomes but only the class into which each falls, we may obtain an approximation to the mean income by taking the mid-value of each class interval to represent all the men in that interval. In this case the mean income is

$$£ \frac{250 \times 8 + 750 \times 15 + 1250 \times 11 + 1750 \times 4 + 2250 \times 2}{8 + 15 + 11 + 4 + 2}$$

or £1012. 10s. Here the mean value is greater than the modal value but we have insufficient information to state the median value.

Consider now the set of examination marks shown in Table 37 taken over two unstreamed and non-selective classes of 32 boys. The sample, albeit small, was fairly representative of the general population, and the questions and marking scheme devised in such a way that the "average child" might expect to obtain about half marks.

TABLE 37

7	29	38	44	48	53	58	67
12	31	39	44	48	54	59	69
15	32	39	45	48	54	59	69
18	32	40	45	49	54	60	71
20	34	41	45	50	54	63	74
21	36	41	46	50	55	64	75
23	37	42	47	51	56	64	79
26	37	43	47	53	57	66	87

Arranging these into class intervals of 10 marks we have the frequency shown in Table 38.

TABLE 38

Marks	Boys	Marks	Boys
0–9	1	50–59	15
10–19	3	60–69	8
20–29	5	70–79	4
30–39	10	80–89	1
40–49	17	90–99	0

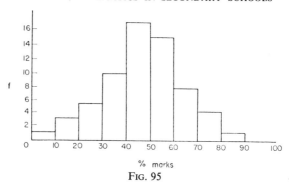
Fig. 95

The corresponding histogram is shown in Fig. 95.

FREQUENCY CURVES AND THE NORMAL DISTRIBUTION CURVE

As the number of readings increases we can reduce the width of the class interval. Had our last sample contained 6400 marks (instead of 64) from a random selection of people over the whole population, we could have taken class intervals of 1 mark instead

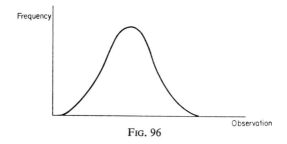
Fig. 96

of 10. The upper bounds of the columns would then have tended to form a curve rather than a series of steps. Indeed, we could plot such results as we plot a frequency polygon, joining the points by a curve instead of straight lines. Such a curve would be called a *frequency curve*.

For a very large number of observations or values the curve would assume the shape shown in Fig. 96.

This is called the *normal distribution curve* and it occurs frequently in social surveys and scientific work. Had the observations been heights of all men over 18 years, the weights of potato

Fig. 97

crop per acre over the whole country or experimental values obtained in the determination of some physical constant, we should have obtained this shape of curve, always assuming that the number of observations or readings was very large.

A common example of the distribution of wear occurs in old stone steps (Fig. 97).

MEAN, MODE, MEDIAN

For some purposes we require a number or quantity which best represents a set of numbers or quantities. The number most commonly used is the *arithmetic mean*. Knowing the average age of a class of children we have not yet met enables us, in some degree, to visualize the whole group. The average speed for a certain journey and the average rainfall of a certain country are measures which, to some extent, typify the whole journey or the climate of the country respectively.

If we have n observations $x_1, x_2, x_3, \ldots, x_n$, the mean value is

$$\frac{x_1 + x_2 \ldots x_n}{n} \quad \text{or} \quad \sum_{r=1}^{n} \frac{x_r}{n}$$

If the observations $x_1, x_2, \ldots, x_n$ occur with frequencies $f_1, f_2, \ldots, f_n$ respectively, then the mean value is

$$\frac{f_1 x_1 + f_2 x_2 \ldots f_n x_n}{f_1 + f_2 \ldots f_n} \quad \text{or} \quad \frac{\sum_{1}^{n} f_r x_r}{\sum_{1}^{n} f_r}$$

or simply

$$\frac{\sum fx}{\sum f}$$

In the set of 64 examination marks quoted in the last section the arithmetic mean is $\frac{3014}{64}$ or 47·1 to three significant figures.

Simple addition of large numbers of readings can be laborious. A quicker way of determining the above mean is to guess the answer and then to calculate the average difference (or deviation) of the values from this guess or *fictitious mean*. Taking 50 as fictitious mean, tabulating the deviations from 50 and deleting equal positive and negative deviations we have Table 39.

TABLE 39

−43	−21	−12	−6	−2	3	8	17
−38	−19	−11	−6	−2	4	9	19
−35	−18	−11	−5	−2	4	9	19
−32	−18	−10	−5	−1	4	10	21
−30	−16	−9	−5	0	4	13	24
−29	−14	−9	−4	0	5	14	25
−27	−13	−8	−3	1	6	14	29
−24	−13	−7	−3	3	7	16	37
−229	−65	−34	−16	−6	12	30	122

The total deviation $164 - 350 = -186$. The average deviation

$$= \frac{-186}{64} = -2\cdot906$$

$\therefore$ mean value $= 50 - 2\cdot9 = 47\cdot1$

to three significant figures.

It is almost self-evident that the average of the deviations is the difference between the true and fictitious means. The proof, if required, is as follows.

Consider a set of elements $x_1, x_2, \ldots, x_n$ occurring with frequency $f_1, f_2, \ldots, f_n$ respectively. Let M be the true mean and A the fictitious mean. Then the average of the deviations

$$= \frac{\sum f(x-A)}{\sum f}$$
$$= \frac{\sum fx}{\sum f} - A$$
$$= M - A$$

The *mode* is the most popular value or the member of the set for which f is greatest. In the example discussed above the mode lies in the range 40–49. It would not be helpful to a manufacturer to know that the mean value of men's shoe sizes in a certain country is 9·37; in this context it would be better to quote the mode or most popular size as 9 or $9\frac{1}{2}$.

The *median* value is that of the middle element when these are ranked in order of magnitude. For a normal distribution the mean, mode and median coincide, but for a skew distribution they are distinct.

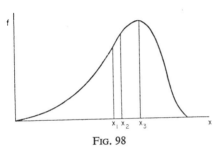

Fig. 98

The mean value (x_1) is the x coordinate of the centroid of the area under the curve (Fig. 98); the median value (x_2) corresponds to the ordinate which bisects the area under the curve and the mode (x_3) is the x coordinate corresponding to the maximum value of f.

There is an approximate relationship between these measures of the form

$$\text{mode} - \text{median} = 2(\text{median} - \text{mean})$$
or
$$x_3 - x_2 = 2(x_2 - x_1)$$

MEASURES OF DISPERSION

In analysing a set of data, the various "mean values" give only a restricted view of the general picture.

The average rainfall of a certain country forms part of the picture of the climate. It does not tell us, however, whether the total rainfall occurred in three torrential downpours, or whether it was spread over a number of rainy days throughout the year. The sets {99, 100, 101} and {1,100,199} have the same mean and median values, but they are clearly very different in respect of the dispersion of their elements.

A more complete picture of these sets is given if we state not only the mean but also the *range* of each as 2 and 198 respectively. This, though better, is still crude; it gives no indication of the distribution of frequencies within the range.

Consider, for example, the distributions shown in Fig. 99.

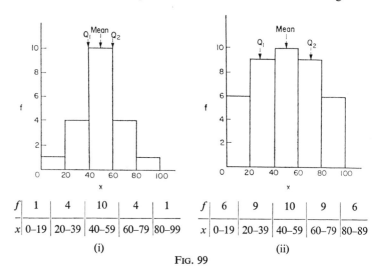

Fig. 99

Distribution (ii) of Fig. 99 has a much wider scatter than (i), and yet the means, modes and ranges are equal.

THE SEMI-INTERQUARTILE RANGE

The median divides the distribution into two equal parts in each of which we have equal numbers of observations. The *quartiles* divide each of these two halves; thus, between the two quartiles where $x = Q_1$, $x = Q_2$ respectively, there lies the middle half of the observations. The *semi-interquartile* range is $\frac{1}{2}(Q_2 - Q_1)$. This clearly gives a better idea of the dispersion than a simple statement of the range.

THE MEAN DEVIATION

The mean deviation about the mean is defined as $\dfrac{\sum f|d|}{\sum f}$, where the values of $|d|$ are simply the numerical sizes, irrespective of sign, of the deviations of the observations or readings from the mean. Thus, in table (i) of Fig. 99 or table (ii) of Fig. 99 we have Table 40.

TABLE 40

x	f	$\|d\|$	$f\|d\|$
10	1	40	40
30	4	20	80
50	10	0	0
70	4	20	80
90	1	40	40

$$\Sigma f|d| = 240$$
$$\Sigma f = 20$$
$$\therefore \text{ mean deviation} = \frac{240}{20} = 12$$

x	f	$\|d\|$	$f\|d\|$
10	6	40	240
30	9	20	180
50	10	0	0
70	9	20	180
90	6	40	240

$$\Sigma f|d| = 840$$
$$\Sigma f = 40$$
$$\therefore \text{ mean deviation} = \frac{840}{40} = 21$$

[The average of the deviations about the true mean is, of course, zero, but by counting all the deviations as positive, i.e. taking $|d|$, we have a measure of the dispersion.]

THE STANDARD DEVIATION

The most widely used measure of dispersion is the standard deviation σ. This is defined as follows: The mean of the squared deviations of a distribution about the mean is called the *variance*. The positive square root of the variance is called the *standard deviation*.

Hence, if the observation x occurring with frequency f has deviation d from the mean, then

$$\sigma = \sqrt{\frac{\sum fd^2}{\sum f}}$$

Using (i) and (ii) of Fig. 99 again we obtain Table 41.

TABLE 41

(i)

x	f	d	d^2	fd^2
10	1	−40	1600	1600
30	4	−20	400	1600
50	10	0	0	0
70	4	20	400	1600
90	1	40	1600	1600

$\Sigma f = 20 \qquad \Sigma fd^2 = 6400$

$$\therefore \sqrt{\frac{\Sigma fd^2}{\Sigma f}} = \sqrt{\frac{6400}{20}} = \sqrt{320}$$
$$\simeq 18$$

(ii)

x	f	d	d^2	fd^2
10	6	−40	1600	9600
30	9	−20	400	3600
50	10	0	0	0
70	9	20	400	3600
90	6	40	1600	9600

$\Sigma f = 40 \qquad \Sigma fd^2 = 26400$

$$\therefore \sqrt{\frac{\Sigma fd^2}{\Sigma f}} = \sqrt{\frac{26400}{40}} = \sqrt{660}$$
$$\simeq 26$$

These give us a measure of the dispersion in each of the two cases.

We have seen that the process of obtaining the mean of readings $x_1, x_2, \ldots, x_n$ with associated frequencies $f_1, f_2, \ldots, f_n$ as $\dfrac{\sum fx}{\sum f}$ is exactly analogous with the process of obtaining the x coordinate of the centroid of particles at $x_1, x_2, \ldots, x_n$ with associated masses $m_1, m_2, \ldots, m_n$ in the form $\dfrac{\sum mx}{\sum m}$.

The process of obtaining the standard deviation about the mean as $\sqrt{\dfrac{\sum fd^2}{\sum f}}$ is also exactly analogous with the process of determining the swing-radius, or radius of gyration, of a set of masses $m_1, m_2, \ldots, m_n$ at distances $x_1, x_2, \ldots, x_n$ from the axis through their centroid in the form

$$Mk^2 = \sum md^2 \quad \text{or} \quad k = \sqrt{\frac{\sum md^2}{\sum m}}$$

SHORT METHOD OF CALCULATING THE STANDARD DEVIATION

In general the mean value M, and hence the values of d, will be decimal quantities. If deviations are taken about the true mean considerable labour is usually involved in the calculation of Σfd^2.

Where the mean value is a whole number this labour is greatly reduced. We now give a method of obtaining σ by taking deviations about a fictitious mean A.

Let the corresponding fictitious standard deviation be s (i.e. s is the value of $\sqrt{\dfrac{\sum fd^2}{\sum f}}$ where the values of d are taken from A, the fictitious mean, instead of M, the true mean). Then for n observations $x_1, x_2, \ldots, x_n$ occurring with frequencies $f_1, f_2, \ldots, f_n$ we have

$$s^2 = \frac{f_1(x_1-A)^2 + f_2(x_2-A)^2 \ldots f_n(x_n-A)^2}{f_1+f_2\ldots f_n}$$

$$\sigma^2 = \frac{f_1(x_1-M)^2 + f_2(x_2-M)^2 \ldots f_n(x_n-M)^2}{f_1+f_2\ldots f_n}$$

$$\therefore s^2 - \sigma^2 = \frac{2(M-A)(f_1x_1+f_2x_2\ldots f_nx_n) + (A^2-M^2)(f_1+f_2\ldots f_n)}{f_1+f_2\ldots f_n}$$

$$= 2(M-A)\frac{\sum fx}{\sum f} + A^2 - M^2$$

$$= 2(M-A)M + A^2 - M^2$$

$$= M^2 - 2AM + A^2$$

$$= (M-A)^2$$

$$\therefore \sigma^2 = s^2 - (M-A)^2$$

From this result it follows that the sum of the squares of the deviations of a set of numbers from any number A is least when A is the mean value of the set. Hence, for example, the moment of inertia of a body is least when taken about an axis passing through its centroid.

Example

Calculate the mean value and standard deviation of the following set of readings.

x	1	2	3	4	5	6	7	8	9	10
f	1	2	4	8	12	11	9	5	3	1

Take as fictitious mean $A = 5$.

TABLE 42

x	f	deviation from $A = 5$	fd	d^2	fd^2
1	1	−4	−4	16	16
2	2	−3	−6	9	18
3	4	−2	−8	4	16
4	8	−1	−8	1	8
5	12	0	0	0	0
6	11	1	11	1	11
7	9	2	18	4	36
8	5	3	15	9	45
9	3	4	12	16	48
10	1	5	5	25	25

$\Sigma f = 56$ $\qquad\qquad\qquad\qquad \Sigma fd = 35 \qquad \Sigma fd^2 = 223$

$\dfrac{\Sigma fd}{\Sigma f} = \dfrac{35}{56} = 0{\cdot}625 \qquad \therefore$ mean $M = 5 + 0{\cdot}625 = 5{\cdot}625$

$\dfrac{\Sigma fd^2}{\Sigma f} = \dfrac{223}{56} = 3{\cdot}982 = s^2 \quad \therefore \sigma^2 = s^2 - 0{\cdot}625^2$

$\qquad\qquad\qquad\qquad\qquad\qquad = 3{\cdot}982 - 0{\cdot}3906$

$\therefore \sigma^2 = 3{\cdot}5914 \qquad\qquad \therefore \sigma = 1{\cdot}895$

standard deviation about the mean 5·625 is 1·90 to 3 significant figures.

THE METHOD OF LEAST SQUARES—CORRELATION

In the previous sections we have discussed the characteristic features of a distribution. The mean and standard deviation of a particular set of observations are sufficient to give us a fair picture of the whole set. In this section we discuss the manner in which equinumerous sets may be compared.

Frequently we meet sets of figures which bear a resemblance to each other. We may find, for example, that the marks obtained by five candidates in a pure mathematics and an applied mathematics examination have certain features in common. It may be that a pupil who gains high marks in one paper also gains high marks in the other; the pupil who gets poor marks in pure mathematics may produce a mediocre paper in applied mathematics. This is not necessarily the case but it often will be so. If the five pupils obtain identical class positions in each of the examinations we should say that there was a high correlation between the sets of marks.

The *Monthly Digest of Statistics* for September 1957 contains the information given in Table 43. Here is a case where we might

TABLE 43

	Supplementary unemployment benefit paid (£1000)	Number of unemployed (1000's)
1951	33	185·8
1952	59	393·5
1953	48	272·7
1954	30	220·1
1955	20	202·2

By kind permission of the Controller, H.M.S.O.

expect a considerable degree of correlation and on examining the figures we see that the benefit paid and the unemployment numbers registered increase and decrease together.

As we shall see later, this degree of "similarity of trend" or correlation is measured on the scale -1 to $+1$. A correlation of $+1$ indicates that two sets of numbers increase or decrease together in exactly the same way, i.e. if y and x are related by the equation $y = mx + c$ (m positive) then x and y are perfectly correlated. Where two sets of observations are totally unrelated in respect of pattern or trends we find zero correlation, and if one set reveals trends which are exactly opposite to those of the other i.e. y increases when x decreases according to $y = -mx + c$ (m positive), then we have a negative correlation of -1.

We begin to look at the question of measuring correlation by considering a problem which frequently occurs in experimental work. Suppose we measure certain values of some quantity y

corresponding to selected values of another quantity x, e.g. effort against load for a certain machine, tension against extension for a given spring, voltage against current through a given resistance, etc. We often find on plotting the graph that the points lie roughly along a straight line. In such a case we draw the best straight line through them. Usually this is simply a matter of judgement, but it can be done quite accurately by calculation.

Consider the following set of results:

x	2	5	7	11	13
y	1	2	4	5	7

Plotting these on a graph we obtain Fig. 100.

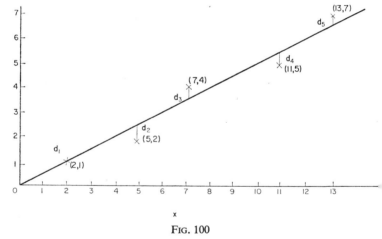

Fig. 100

Assuming that the values of x are correct, then the best straight line $y = mx$ is that for which the sum of the squares of the deviations of the y values is least. Denoting this sum by S, and noting that $d_1 = (2m-1)$, $d_2 = (5m-2)$, $d_3 = (4-7m)$, $d_4 = (11m-5)$, $d_5 = (7-13m)$, we have

$$S = (2m-1)^2 + (5m-2)^2 + (7m-4)^2 + (11m-5)^2 + (13m-7)^2$$

For a minimum value of S,

$$\frac{dS}{dm} = 0 \text{ and } \frac{d^2S}{dm^2} \text{ is positive,}$$

$\therefore\ 4(2m-1) + 10(5m-2) + 14(7m-4) +$
$\qquad\qquad\qquad 22(11m-5) + 26(13m-7) = 0$

$\therefore\ m = \dfrac{372}{736} = 0\cdot505$ or $0\cdot51$ to 2 decimal places.

Hence the best straight line is $y = 0\cdot51x$.

In the above example we have taken the simplified case in which the best straight line is of the form $y = mx$. In general a linear relationship between two variables takes the form $y = mx + c$. If $(x_1 y_1)(x_2 y_2) \ldots (x_r y_r)$ are points on this line, then $(\bar{x}, \bar{y})$, where $\bar{x}$ is the mean of the x values and $\bar{y}$ is the mean of the y values, must also lie on the line. Transferring the origin of the coordinates to $(\bar{x}, \bar{y})$ by the transformation

$$X = x - \bar{x}$$
$$Y = y - \bar{y}$$

we have $\qquad Y + \bar{y} = m(X + \bar{x}) + c$

but $\qquad\qquad \bar{y} = m\bar{x} + c$

$\qquad \therefore\ Y = mX$

Hence, in a case where the values of x and y indicate a law of the form $y = mx + c$, we tabulate values of $x - \bar{x}$, (X), $y - \bar{y}$, (Y) (i.e. the deviations of the values from their respective means), and obtain the value of m in $Y = mX$ for which the variance of the Y values is a minimum.

Suppose that we now have a set of values of x and y which have been transformed in this way.

x	x_1	x_2	x_3	...	x_n
y	y_1	y_2	y_3	...	y_n

Assume that the x values are correct and that the best straight line is $y = mx$. Proceeding as before we have

$$S = (mx_1 - y_1)^2 + (mx_2 - y_2)^2 \ldots (mx_n - y_n)^2$$

For minimum S,

$$\frac{dS}{dm} = 0 \quad \text{and} \quad \frac{d^2S}{dm^2} \text{ is positive.}$$

Differentiating we have

$$2x_1(mx_1-y_1)+2x_2(mx_2-y_2)\ldots 2x_n(mx_n-y_n) = 0$$

$$\therefore m = \frac{x_1y_1+x_2y_2\ldots x_ny_n}{x_1^2+x_2^2\ldots x_n^2} = \frac{\sum x_r y_r}{\sum x_r^2}$$

Hence the required equation may be written in the form

$$y = \frac{\sum x_r y_r}{\sum x_r^2} \cdot x$$

or

$$\frac{y}{\sqrt{\frac{\sum y_r^2}{n}}} = \frac{\sum x_r y_r}{\sqrt{\sum x_r^2 \sum y_r^2}} \cdot \frac{x}{\sqrt{\frac{\sum x_r^2}{n}}}$$

or

$$\frac{y}{\sigma_y} = r \cdot \frac{x}{\sigma_x}$$

where

$$r = \frac{\sum x_r y_r}{\sqrt{\sum x_r^2 \sum y_r^2}}$$

σ_x, σ_y are the standard deviations of the x and y values and r is called the *coefficient of correlation* between the two variables. r is therefore a number which measures the extent of the change occurring in y (expressed as a proportion of its standard deviation) for a corresponding change in x (measured in terms of its standard deviation).

We can show that this is a very convenient measure of correlation with reference to three examples.

Example 1

x	1	2	3	4	5
y	4	6	8	10	12

The mean values of x and y are 3 and 8 respectively. Let the deviations of the x,y values from their respective means be X, Y (Fig. 101).

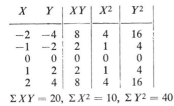
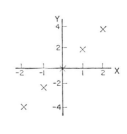

X	Y	XY	X^2	Y^2
-2	-4	8	4	16
-1	-2	2	1	4
0	0	0	0	0
1	2	2	1	4
2	4	8	4	16

$\Sigma XY = 20$, $\Sigma X^2 = 10$, $\Sigma Y^2 = 40$

FIG. 101

$$\therefore r = \frac{\Sigma XY}{\sqrt{\Sigma X^2 \cdot \Sigma Y^2}} = \frac{20}{\sqrt{400}} = 1$$

This is a case of perfect correlation, for x,y are co-related or related exactly by the linear law $y = 2x+2$. They increase together in a steady, precise manner.

Example 2

Had we taken

x	1	2	3	4	5
y	0	-2	-4	-6	-8

the mean values would have been 3 and -4 respectively, and our table of deviations would have become as Fig. 102.

X	Y	XY	X^2	Y^2
-2	4	-8	4	16
-1	2	-2	1	4
0	0	0	0	0
1	-2	-2	1	4
2	-4	-8	4	16

$\Sigma XY = -20$, $\Sigma X^2 = 10$, $\Sigma Y^2 = 40$

FIG. 102

In this case $r = \dfrac{-20}{\sqrt{400}} = -1$

This is a case of perfectly negative correlation ($y = -2x+2$). As x increases y decreases in the same steady manner.

Example 3

Consider now the table of values

x	1	2	5	2	5
y	1	4	3	-2	-1

Here, the mean values of x and y are 3 and 1 respectively. The table of deviations is then as in Table 44.

TABLE 44

X	Y	XY	X^2	Y^2
-2	0	0	4	0
-1	3	-3	1	9
2	2	4	4	4
-1	-3	3	1	9
2	-2	-4	4	4

and $r = \dfrac{\sum XY}{\sqrt{\sum X^2 \cdot \sum Y^2}} = 0$

i.e. there is no correlation at all between the two sets of values. Representing the points graphically we see that they are quite haphazardly situated and exhibit no discernible trend or co-relationship (Fig. 103).

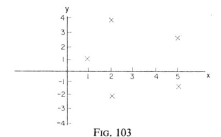

Fig. 103

Example 4

Marks obtained by five students in spot tests in mathematics and physics were as follows:

Maths	10	8	4	7	6
Physics	9	8	3	8	7

If we examine these we see that there is a fair degree of correlation. We might expect quite a high value of r. Denoting the deviations by x_m, x_p, and noting that the mean in each case is 7 we have Table 45.

TABLE 45

x_m	x_p	$x_m x_p$	x_m^2	x_p^2
3	2	6	9	4
1	1	1	1	1
−3	−4	12	9	16
0	1	0	0	1
−1	0	0	1	0
		19	20	22

$$\therefore r = \frac{\sum x_m x_p}{\sqrt{\sum x_m^2 \cdot \sum x_p^2}} = \frac{19}{\sqrt{20 \times 22}} = 0.91$$

SIGNIFICANCE

A high coefficient of correlation between two sets of values does not necessarily indicate a connexion or that a strong causal factor is involved.

For example if there is a high correlation between smoking and the incidence of lung cancer, the fact is certainly worth investigating. We may not conclude from this alone, however, that smoking *causes* lung cancer. There may, in fact, be a linking factor. For example, bad weather (C) may *cause* low attendances at the Test Match (A) and light traffic on the coast roads (B) but we cannot say that A *causes* B. If A and B separately correlate highly with

C then, of course, they will correlate with each other without being connected in any other way.

Indeed, there may be no linking factor at all. Any two quantities which increase together will yield a positive correlation coefficient without having the remotest connexion with each other.

It is therefore necessary to determine whether r, when evaluated, has any *significance*. If a dependence between two quantities is known to exist the correlation coefficient may be used to measure it, but a correlation coefficient can never be used to *establish* a dependence or causal link.

Generally, if the value of r for two sets could be obtained by pure chance at least one in ten times, then we take r to have no significance.

THE NORMAL DISTRIBUTION CURVE

We have seen that certain features such as height, weight, shoe size, intelligence, etc., are distributed in a very symmetrical way over the whole population, and that the frequency curve is of the form shown in Fig. 104. This curve is known as the normal distribution curve or, for some purposes, as the Gaussian frequency curve.

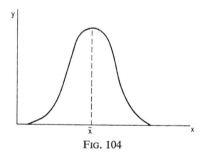

Fig. 104

If we label the frequency axis y, the observation axis x, it can be proved that the equation of the curve is

$$y = \frac{1}{\sigma\sqrt{2\pi}}\, e^{-(x-\bar{x})^2/2\sigma^2}$$

where $\bar{x}$ is the mean value of the observations and σ is their standard deviation about $\bar{x}$.

If we now transfer the y axis to the line $x = \bar{x}$, we obtain the more symmetrical graph shown in Fig. 105.

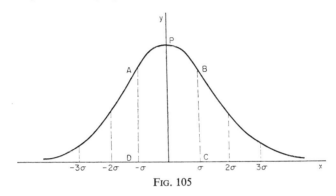

Fig. 105

Its equation is now $y = \dfrac{1}{\sigma\sqrt{2\pi}}\, e^{-x^2/2\sigma^2}$

Any graph of the form $y = C\, e^{-x^2}$ has the shape of the normal distribution curve. We select the coefficient $\dfrac{1}{\sigma\sqrt{2\pi}}$ since it can be shown that the area under the curve $y = \dfrac{1}{\sigma\sqrt{2\pi}}\, e^{-x^2/2\sigma^2}$ is unity.

Gauss proved that the probability of an element of a normal distribution having a deviation from the mean lying between x and $x+\delta x$, is $\dfrac{1}{\sigma\sqrt{2\pi}} \cdot e^{-x^2/2\sigma^2} \cdot \delta x$. The probability that an element has a deviation somewhere between $-\infty$ and $+\infty$ is

therefore
$$\int_{-\infty}^{\infty} \dfrac{e^{-x^2/2\sigma^2}}{\sigma\sqrt{2\pi}}\, dx$$

This, of course, is the area under the curve, i.e. unity. But it is an absolute certainty that any element has a deviation in this range and for absolute certainty the probability is unity. Hence, by selecting the coefficient $\dfrac{1}{\sigma\sqrt{2\pi}}$, we arrange that the probability of an element having deviation between $\pm d$ is numerically equal to the area under the curve between $x = \pm d$.

The probability that an element has deviation within the range $\pm \sigma$ is thus numerically equal to the area $APBCD$, and, of course, the probability is that fraction of the whole distribution which lies in this range. Table 46 may be of interest.

TABLE 46

Range	Probability that an element lies in this range	Percentage of the whole distribution within this range
$-\frac{1}{4}\sigma$ to $\frac{1}{4}\sigma$	·198	19·8
$-\frac{1}{2}\sigma$ to $\frac{1}{2}\sigma$	·383	38·3
$-\sigma$ to σ	·683	68·3
$-\frac{3}{2}\sigma$ to $\frac{3}{2}\sigma$	·866	86·6
-2σ to 2σ	·954	95·4
-3σ to 3σ	·997	99·7

Another feature of interest about the normal curve concerns the position of its points of inflexion.

Since
$$y = \frac{1}{\sigma\sqrt{2\pi}} \exp{-x^2/2\sigma^2}$$

we have
$$\frac{dy}{dx} = \frac{1}{\sigma\sqrt{2\pi}} \left[\frac{-2x}{2\sigma^2} \cdot \exp{-x^2/2\sigma^2} \right]$$

and
$$\frac{d^2y}{dx^2} = \frac{-1}{\sigma^3\sqrt{2\pi}} \left[x \cdot \frac{-2x}{2\sigma^2} \exp{-x^2/2\sigma^2} + \exp{-x^2/2\sigma^2} \right]$$

At the points of inflexion $\dfrac{d^2y}{dx^2} = 0$, i.e. $x = \pm \sigma$.

Hence A and B are the points of inflexion.

In the present chapter we have outlined some of the basic ideas of the study of statistics and, perforce, omitted a very great deal.

The ideas of mean and dispersion are not only simple but important; they should form part of any O-level course in mathematics.

In the sixth form, certain pupils taking the subject to A-level will go well beyond the point we have reached in this chapter. Pupils studying economics or social studies need to learn something of index numbers, moving averages, the operation of insurance companies and the correlation of financial trends. Those reading science and engineering will obviously benefit from a clear understanding of correlation and significance, Gauss's law of error and, perhaps, the elements of quality control.

Much, of course, will depend on the individual teacher, the time available and the ability of the pupils, but no child should leave school completely unaware that patterns exist in certain distributions and that facts can be made to emerge from figures.

Useful Reference Books

ADLER, I., *Probability and Statistics for Everyman*, Dennis Dobson.
ALLENDOERFER and OAKLEY, *Principles of Mathematics*, McGraw-Hill.
BROOKES, B. C. and DICK, W. F. L., *Introduction to Statistical Method*, Heinemann.
CONNOLLY, T. G. and SLUCKIN, W., *Statistics for the Social Sciences*, Cleaverhulme.
HAYS, S., *Outline of Statistics*, Longmans.
LEVY and PREIDEL, *Elementary Statistics*, Nelson.
MCINTOSH, D. M., *Statistics for the Teacher*, Pergamon Press.
MORONY, M. J., *Facts from Figures*, Pelican.
MOUNSEY, J., *Introduction to Statistical Calculations*, English Universities Press.
PICKARD, R. E., *Statistics*, Cassell.
SHERLOCK, A. J., *Probability and Statistics*, E. Arnold.
VESSELO, I. R., *How to Read Statistics*, Harrap.
The Teaching of Statistics, Mathematics Teaching.

8
PATTERNS IN ARITHMETIC

THE PRESENT SITUATION

Anyone who has worked and played with young children will have noticed their natural fascination with numbers and number patterns. Many teachers at the end of term, the serious business of examinations behind them, have abandoned "the syllabus" and brought out a private stock of number acrostics, magic squares, fallacies and puzzles in order to maintain the interest and enthusiasm which would otherwise flag. Indeed, interest in such things is not confined to children. Brain twisters, puzzle corners and mathematical recreations appear in national publications, and the mathematics teacher who has not been confronted by a paradox of some kind sent in by a pupil's father is a rare bird indeed.

Many of these surprising number facts or unsuspected patterns are of real mathematical interest and it seems a pity that they are not exploited more in the normal work of the classroom.

One reason is that many examination questions in arithmetic, disguised though they are as problems on income tax, rates and taxes, stocks and shares, public expenditure or mensuration, consist essentially of tedious applications of the four rules. Other questions demand exact answers or far more accurate ones than would be required in a real-life situation. By the time a pupil has had sufficient practice in performing calculations which slide rules and desk calculators do so much more quickly, there is little time left for topics of intrinsic interest.

Of course, "civic arithmetic" has its proper place in a child's education, and counting being one of the basic skills by which we organize and convey our impressions of our environment, it is essential that children should understand numbers. Furthermore, multiplication tables up to the "12 times" should be known by the time the secondary stage is reached.

What we should do then is to frame the work in such a way that, using his skills, the pupil discovers more patterns and gains deeper understanding of the work already done; too often these skills are practised via harder additions and multiplications of the same type.

The processes by which we add together, or multiply together, the numbers 235 and 123 in the denary scale are, in fact, very sophisticated ones. When adding, for example, we do not actually add the numbers together; we take advantage of the place system and simply add the respective digits. The child who really understands this idea of place value is unlikely to have difficulty with the position of decimal points and can soon calculate in other scales of notation.

It is often a good idea to revise these processes at the beginning of the secondary stage by computation in other scales of notation, and it is something new. Mansfield and Thompson do just this in their series of textbooks *Mathematics—A New Approach*.

At the same time it should be pointed out that some work of this kind is already being done in primary schools, and that some excellent books and material have become available in recent years. The Dienes Multibase Arithmetic Blocks (bases 3, 4, 5, 6 and 10) and Algebraical Experience Materials have been introduced into some Leicestershire primary schools. In other schools the Stern Apparatus, Colour Factor Apparatus and Cuisenaire material are used. Also, experiments with desk calculating machines have been carried out. The children not only exercise their skills unbidden in checking the machine, but the method of operating a desk calculator calls for careful thought and an understanding of place-value.

Books too are available, and an extraordinarily attractive series of primary and junior books has just been published by Addison-Wesley. In these arithmetic is developed through the notion of sets. Starting with the basic ideas of relative size and order, and mainly pictorially. the concepts of inequality, one-to-one correspondence and equivalence of sets are gradually built up. From pictorial sets are abstracted the corresponding ideas in ordinal and cardinal numbers and thence the operations of addition, subtraction and multiplication.

Finally one should mention the *Mathematical Gazette*, *Mathematics Teaching*, and a new publication, *Teaching Arithmetic* (Pergamon). All of these journals contain interesting and lively articles on modern approaches in the classroom.

Reverting to the secondary school situation, it is worth noting that whereas our existing O-level syllabuses contain a good deal of heavy arithmetic, A-level syllabuses contain virtually none. In this respect we go to the other extreme and we accept, for example, answers which state that an angle of friction is $\tan^{-1} \frac{\sqrt{3}+1}{2\sqrt{3}}$ or that a certain area is $\frac{1}{10}(2\sqrt{2}\, e^{3\pi/4} - 3)$ units. One would not expect such results to be evaluated in every case, but pupils should realize that in research departments of industry numerical answers are required. Furthermore, the great majority of integrals which arise in actual practice can only be evaluated by approximate methods. A little practice with a desk calculator computing such results is often time well spent.

In the following sections we discuss some of the interesting patterns and properties to be found in numbers. Many of these are capable of investigation and "discovery" by quite young children. Many of them lead to fundamental ideas.

PATTERNS IN ADDITION AND MULTIPLICATION

The first issue of *Teaching Arithmetic* (Pergamon Press) contains an interesting article on the "9 times" table, shown as Table 47.

TABLE 47

$1 \times 9 = 9$ $7 \times 9 = 63$
$2 \times 9 = 18$ $8 \times 9 = 72$
$3 \times 9 = 27$ $9 \times 9 = 81$
$4 \times 9 = 36$ $10 \times 9 = 90$
$5 \times 9 = 45$ $11 \times 9 = 99$
$6 \times 9 = 54$ $12 \times 9 = 108$

We observe certain patterns of which the following are a few:

(a) The sum of the digits of any product is 9.

$8 \times 9 = 72, \quad 7+2 = 9; \quad 23 \times 9 = 207, \quad 2+0+7 = 9$, etc.

This suggests that any number, the sum of whose digits is 9, is itself divisible by 9.

Test 24768; $\quad 2+4+7+6+8 = 27, \quad 2+7 = 9$

Hence 24768 is divisible by 9 and, in fact, is 2752×9.

(b) At 6×9 the digits of the product are reversed.

$$5 \times 9 = 45 \qquad 4 \times 9 = 36$$
$$6 \times 9 = 54 \qquad 7 \times 9 = 63$$

(c) Take any number and divide it by 9.

$$\begin{array}{r} 8538 \\ \hline 9)\overline{76843} \end{array} \quad \text{remainder 1}$$

We now note that $7+6+8+4+3 = 28;\ 2+8 = 10;\ 1+0 = 1$ and 1 is the remainder in the division.

(d) Take any number, permute the digits in any fashion and subtract.

$$\begin{array}{r} 76843\,- \\ 34876 \\ \hline 41967 \end{array}$$

We then have $4+1+9+6+7 = 27;\ 2+7 = 9$. In the *Mathematical Gazette*, May 1947, Parameswaran showed that if a, b are two groups each of k digits, then

$ab - ba = (a-b).(9)_k \quad \text{where} \quad (9)_k = 999 \ldots \text{ to } k \text{ digits}$
e.g. $\quad 72 - 27 = (7-2) \times 9$
$2422 - 2224 = (24-22) \times 99$
$123122 - 122123 = (123-122) \times 999$

This is extremely interesting and one might well inquire why it should be so. Has it anything to do with the fact that 9 is one less than the radix of the scale of notation? In the scale of 7 should we find a persistence of sixes? We investigate in Table 48.

TABLE 48

$$
\begin{array}{ll}
1 \times 6 = 6 & 0 + 6 = 6 \\
2 \times 6 = 15 & 1 + 5 = 6 \\
3 \times 6 = 24 & 2 + 4 = 6 \\
4 \times 6 = 33 & 3 + 3 = 6 \\
5 \times 6 = 42 & 4 + 2 = 6 \\
6 \times 6 = 51 & 5 + 1 = 6
\end{array}
$$

Further, the digits of the products after 4×6 are reversed.

$$
\begin{array}{ll}
3 \times 6 = 24 & 2 \times 6 = 15 \\
5 \times 6 = 42 & 6 \times 6 = 51
\end{array}
$$

What about $ab - ba = (a-b).(n-1)_k$ base n? Let us try $2422 - 2224$. In the scale of 7 this gives

$$165 \quad \text{or} \quad 2 \times 66$$

i.e. $\quad 2422 - 2224 = (24-22).66 \quad \text{(base 7)}$

Again, in the scale of 5,

$$2422 - 2224 = 143 = (24-22).44 \quad \text{(base 5)}$$

The reader may care to take the matter further and show tha these patterns, which at first sight appear peculiar to the figure 9 are, in fact, repeated for the digit $(k-1)$ in the arithmetic to base k. Better still, let the pupils do it.

CHINESE MULTIPLICATION

An amusing variation of the normal method of setting out long multiplication consists of writing multiplicand and multiplier along the top and down the right-hand side of a 3×3 matrix. Corresponding partial products are entered in each square as shown and the final result is obtained by summing the elements in each diagonal, carrying tens and hundreds where necessary to the next diagonal.

Thus, for 56×43 the product is 2408 (Fig. 106), and $235 \times 123 = 28905$ (Fig. 107).

PATTERNS IN ARITHMETIC

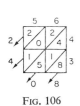
Fig. 106

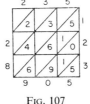
Fig. 107

Fig. 108

To check that we are adding together digits of the same place value we consider the partial products of two numbers each containing tens (T) and units (U) (Fig. 108).

NUMBERS OF THE FORM $6N+1$

If a and b are positive integers we have

$$(6a+1)(6b+1) = 6[6ab+a+b]+1$$
$$(6a-1)(6b-1) = 6[6ab-(a+b)]+1$$

and $6ab \pm (a+b)$ are integers (say N). Thus $6N+1$, unless prime, has a pair of factors of the form $6n+1$ or $6n-1$.

e.g. if $n = 4$ $25 = 5 \times 5$ $n = 9$ $55 = 5 \times 11$
         $n = 8$ $49 = 7 \times 7$

but note,

if $N = 2$ 13 is prime $N = 7$ 43 is prime
     $N = 3$ 19 is prime $N = 11$ 67 is prime
     $N = 5$ 31 is prime

Is $6N+1$ prime for N prime? We find that it is not, but the pattern is worth pursuing.

MAGIC SQUARES

An amusing way of revising, say, mental addition is the construction of squares of distinct numbers in such a way that the sum of any row, diagonal or column is a constant quantity.

These are called magic squares. Numerous examples of these occur in books on mathematical puzzles and recreations. An extremely simple example is

$$\begin{array}{ccc} 12 & 5 & 10 \\ 7 & 9 & 11 \\ 8 & 13 & 6 \end{array}$$

in which the sum of any row, column or diagonal is 27. Mr. D. C. Cross, writing in the *Mathematical Gazette*, October 1961, gives a large number of examples of magic squares which possess even more properties. For example the square

$$\begin{array}{ccc} 2 & 7 & 6 \\ 9 & 5 & 1 \\ 4 & 3 & 8 \end{array}$$

is not only magic, but the sum of the squares of the elements on the opposite sides is also equal. Thus,

$$2^2 + 7^2 + 6^2 = 4^2 + 3^2 + 8^2 = 89$$
and
$$2^2 + 9^2 + 4^2 = 6^2 + 1^2 + 8^2 = 101$$

Furthermore there is a pattern in this which is worth investigation.

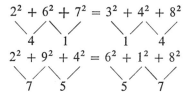

Dr. C. Dudley Langford has written frequently in the *Mathematical Gazette* on the subject and many of his creations reveal him to be a numerical manipulator *par excellence*. In the "super" magic square shown below the "magic" number 34 is the sum of (a) each row, (b) each column, (c) each diagonal, (d) two broken diagonals, (e) the centre four elements, (f) the corner elements,

(g) the centre pairs of opposite sides and (h) the corner elements of any three by three square.

$$\begin{array}{cccc} 1 & 7 & 10 & 16 \\ 14 & 12 & 5 & 3 \\ 15 & 9 & 8 & 2 \\ 4 & 6 & 11 & 13 \end{array}$$

(*Mathematical Gazette*, May 1956)

There is no general method of constructing magic squares (Fig. 109) but rather there exists a number of special methods. The magic square of order one is, of course, a trivial case and no magic squares of order two exist. For any square of odd order k

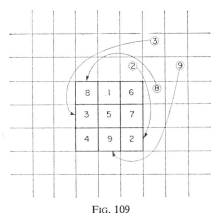

Fig. 109

higher than this, however, we may proceed as follows. Write 1 in the middle of the top row and then fill in, in order, the spaces in the upward broken diagonal through 1 with 2, 3, ... up to k. When this diagonal has been completed we write $k+1$ in the space below k and proceed as before until the whole square is completed.

It is not necessary to commence with 1, it is sufficient to proceed in arithmetical progression with any common difference d, and initial term a.

$a + 7d$	a	$a + 5d$
$a + 2d$	$a + 4d$	$a + 6d$
$a + 3d$	$a + 8d$	$a + d$

thus, for $a = 5, d = 3$ we have

$$\begin{array}{ccc} 26 & 5 & 20 \\ 11 & 17 & 23 \\ 14 & 29 & 8 \end{array}$$

Alternatively we can commence at the space above the middle of the square, fill the upward diagonal and start each new diagonal two spaces above the last entry.

There are many other methods of constructing these squares and a detailed account of the whole subject is found in M. Kraitchik's book *Mathematical Recreations*.

MAGIC MATRICES

In the *Mathematical Gazette* of May 1955, A. D. and K. H. V. Booth proved another extraordinary property of magic squares. They showed that the inverse of a matrix consisting of elements of a magic square is itself a magic square whose row and column sums are the reciprocal of those of the original square. As an illustration consider the matrix

$$a_{rs} = \begin{pmatrix} 8 & 1 & 6 \\ 3 & 5 & 7 \\ 4 & 9 & 2 \end{pmatrix}$$

Now $A_{sr} = \begin{pmatrix} -53 & 52 & -23 \\ 22 & -8 & -38 \\ 7 & -68 & 37 \end{pmatrix}$ and $|a_{rs}| = -360$

$\therefore$ the inverse of a_{rs} is $\dfrac{1}{360} \begin{pmatrix} 53 & -52 & 23 \\ -22 & 8 & 38 \\ -7 & 68 & -37 \end{pmatrix}$

The elements of a_{rs} form the magic square previously considered with column, row and diagonal sum 15. The inverse matrix on inspection also has elements which form a magic square with column, row and diagonal sum $\dfrac{1}{15}$. The reader may like to verify that the product of two magic matrices of equal order, is a "magic" matrix in respect of its rows and columns.

LATIN SQUARES

A latin square of order k is formed with k sets of the numbers 1 to k arranged in such a way that no row or column contains the same number twice. Thus, using three sets of the numbers 1, 2, 3 we form the latin square:

$$\begin{array}{ccc} 1 & 3 & 2 \\ 3 & 2 & 1 \\ 2 & 1 & 3 \end{array}$$

Notice that in this case the sum of the elements in any row, column or diagonal is constant. (This is not a magic square of course as its elements are not all distinct.) However, we need not necessarily use numbers; the following are both examples of latin squares.

$$\begin{array}{ccc} A & B & C \\ B & C & A \\ C & A & B \end{array} \qquad \begin{array}{ccc} 1 & \omega & \omega^2 \\ \omega & \omega^2 & 1 \\ \omega^2 & 1 & \omega \end{array}$$

We remember the second of these arrangements from the multiplication table of the group of rotations of the equilateral triangle. Further, all these patterns are isomorphic under the one to one correspondences.

$$1 \leftrightarrow A \leftrightarrow 1$$
$$3 \leftrightarrow B \leftrightarrow \omega$$
$$2 \leftrightarrow C \leftrightarrow \omega^2$$

Trial and error will soon convince the reader that there is, in fact, only one latin square of order 3. It might appear that

$$\begin{array}{ccc} C & A & B \\ A & B & C \\ B & C & A \end{array}$$

is an alternative form, but on examination it is simply a permutation

$$\begin{pmatrix} A & B & C \\ C & A & B \end{pmatrix}$$

of the elements.

For latin squares of order greater than three the number of distinct patterns increases rapidly. One of the latin squares formed of the elements $I\ A\ B\ C$ may be written in the form of a group multiplication table. Associative and closure properties hold. I is the identity element and to every element there corresponds one, and only one, inverse.

	I	A	B	C
I	I	A	B	C
A	A	B	C	I
B	B	C	I	A
C	C	I	A	B

The reader may care to check that this is isomorphic with the group $\{1, i, -1, -i\}$ under multiplication; the group of matrices

$$\left\{ \begin{pmatrix} 1 & 0 \\ 0 & 1 \end{pmatrix};\ \begin{pmatrix} 0 & 1 \\ -1 & 0 \end{pmatrix};\ \begin{pmatrix} -1 & 0 \\ 0 & -1 \end{pmatrix};\ \begin{pmatrix} 0 & -1 \\ 1 & 0 \end{pmatrix} \right\}$$

under matrix multiplication and the integers $\{1, 2, 4, 3\}$ mod 5 under multiplication and reduction.

Finally, if we take two of the other latin squares of order four,

A	I	B	C
C	B	I	A
I	A	C	B
B	C	A	I

a	c	i	b
i	b	a	c
b	i	c	a
c	a	b	i

and add together corresponding elements (in the same way that we add together matrices), we obtain the latin square

$A+a$	$I+c$	$B+i$	$C+b$
$C+i$	$B+b$	$I+a$	$A+c$
$I+b$	$A+i$	$C+c$	$B+a$
$B+c$	$C+a$	$A+b$	$I+i$

PATTERNS IN ARITHMETIC 195

which is a magic square for all values of the letters! This constitutes one way of forming magic squares of the fourth order.

TRIANGULAR NUMBERS (Fig. 110)

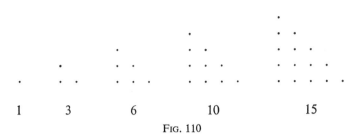

FIG. 110

The triangular numbers are of great interest and they provide a starting point for several topics. Pyramids of cans in grocers' shop windows provide a commonplace example. Mathematically they are commonly associated with the binomial expansion

$$(1-x)^{-3} = 1+3x+6x^2+10x^3+15x^4+21x^5 \ldots$$

The coefficients also appear as one of the oblique rows of Pascal's triangle (from which, of course, the coefficients of the expansions $(1-x)^{-n}$ for all positive integral n may be read off) (Fig. 111).

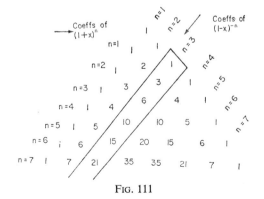

FIG. 111

They may also be written as progressions of natural numbers:

$$1 = 1$$
$$3 = 1+2$$
$$6 = 1+2+3$$
$$10 = 1+2+3+4$$

Determination of the size of the 10th (T_{10}) or nth (T_n) triangular number leads us straight into the realm of arithmetical progressions, and, of course,

$$T_n = \sum_{r=1}^{n} r = \tfrac{1}{2}n(n+1)$$

Further, the sum of any two consecutive triangular numbers is a square number:

$$1+3 = 2^2$$
$$3+6 = 3^2$$
$$6+10 = 4^2$$
$$\cdots$$
$$T_{n-1}+T_n = \tfrac{1}{2}(n-1)n + \tfrac{1}{2}n(n+1) = n^2$$

Another relationship is $T_n + T_m + mn = T_{m+n}$.

For
$$\frac{n(n+1)}{2} + \frac{m(m+1)}{2} + mn = \frac{m^2 + 2mn + n^2 + m + n}{2}$$
$$= \frac{(m+n)(m+n+1)}{2}$$
$$= T_{m+n}$$

This leads to a whole set of interesting patterns. If $mn = 6$ we have either $m = 6\ n = 1$, or $m = 3\ n = 2$.

Hence $T_6 + T_1 + 6 = T_7$
and $T_3 + T_2 + 6 = T_5$
$\therefore T_7 - T_6 - T_1 = T_5 - T_3 - T_2$
or $T_7 + T_3 + T_2 = T_6 + T_5 + T_1$

An unlimited set of relationships of this form may be discovered. Writing in the *Mathematical Gazette*, October 1960, M. N. Khatri develops this idea and shows that there exists a fascinating set of relationships between T_2, T_4 and T_5:

(i) $T_2+T_4+T_5$ is a triangular number. (Actually it is T_7)
(ii) The product of any two plus the third is a triangular number.
(iii) $\Sigma T_2 T_4$ (i.e. $T_2 T_4 + T_2 T_5 + T_4 T_5$) is a square number ($15^2$).
(iv) $2 \Sigma T_2 T_4 = T_2 T_4 T_5$.

For those who are interested, the topic is developed much further in the same journal by Venkatacholam Iyer (February 1962) and G. Stanilov (October 1962).

SQUARE NUMBERS

The sum of the squares of the natural numbers is required in many situations. By summing the identities

$$n^3 - (n-1)^3 = 3n^2 - 3n + 1$$
$$(n-1)^3 - (n-2)^3 = 3(n-1)^2 - 3(n-1) + 1$$
$$\cdot \quad \cdot \quad \cdot \quad \cdot \quad \cdot \quad \cdot$$
$$2^3 - 1^3 = 3.2^2 - 3.2 + 1$$
$$1^3 - 0^3 = 3.1^2 - 3.1 + 1$$

we obtain $\quad 3 \sum n^2 = n^3 + 3 \sum n - n$

or $\quad \sum n^2 = \tfrac{1}{6} n(n+1)(2n+1)$

The number of tennis balls in a square pyramid 6 layers deep is

$$1 + 2^2 + 3^2 \ldots 6^2 \quad \text{or} \quad \sum_{1}^{6} n^2 \quad \text{or} \quad 91.$$

The area under the curve $y = x^2$ from 0 to x may be approximated to a series of rectangular areas $\dfrac{x}{n}$ wide and $\left(\dfrac{rx}{n}\right)^2$ high in the case of the rth column. The total area

$$\simeq \frac{x^3}{n^3} \sum_{r=1}^{n} r^2 \quad \text{or} \quad \frac{x^3}{3} + \frac{x^3}{2n} + \frac{x^3}{6n^2}$$

As n tends to infinity this value tends to the exact area $x^3/3$.

We may begin to introduce basic ideas in the calculus by considering the pattern of differences between cubes and squares:

```
1    8    27   64   125        1    4    9    16   25
   7    19   37   61              3    5    7    9
     12   18   24                   2    2    2
        6    6                        0    0
           0
```

The cubes themselves show an interesting pattern which involves the squares of the triangular numbers:

$$1^3 = 1^2$$
$$1^3 + 2^3 = 3^2$$
$$1^3 + 2^3 + 3^3 = 6^2$$
$$1^3 + 2^3 + 3^3 + 4^3 = 10^2$$

This pattern continues indefinitely.

$$\sum_{r=1}^{n} r^3 = [T_n]^2 = \left[\frac{n(n+1)}{2}\right]^2$$

and we have a formula for the sum of the cubes of the natural numbers.

Many of the prime numbers may be expressed as the sum of two squares. For example we have:

$$2 = 1^2 + 1^2 \qquad 41 = 5^2 + 4^2$$
$$5 = 2^2 + 1^2 \qquad 61 = 6^2 + 5^2$$
$$13 = 3^2 + 2^2 \qquad 113 = 8^2 + 7^2$$
$$17 = 4^2 + 1^2$$

A class of juniors will need little prompting to see whether they can find any more.

Finally, there are some extremely interesting runs of squares which can be discovered using quite elementary algebra. In the *Mathematical Gazette*, December 1961, T. H. Beldon described one way of generalizing the well-known result $3^2 + 4^2 = 5^2$. Here we have two consecutive square numbers equal to the square of the next number. Can we find three consecutive square numbers which are equal to the sum of the next two consecutive square numbers, i.e. can we solve

$$(n-1)^2 + n^2 + (n+1)^2 = (n+2)^2 + (n+3)^2 ?$$

PATTERNS IN ARITHMETIC 199

We find that $n = 11$ (or -1) and obtain the next set

$$10^2 + 11^2 + 12^2 = 13^2 + 14^2$$

Proceeding in this way we find:

$$21^2 + 22^2 + 23^2 + 24^2 = 25^2 + 26^2 + 27^2$$
$$36^2 + 37^2 + 38^2 + 39^2 + 40^2 = 41^2 + 42^2 + 43^2 + 44^2$$

Notice that these runs start with the successive "even" triangular numbers $3, (T_2)$ $10, (T_4)$ $21, (T_6)$ $36, (T_8)$. Can we generalize the result still further and always find a run starting with (T_{2n})?

PRIME NUMBERS

A useful exercise in mental division is to use the "sieve of Eratosthenes" to find all the primes less than 200. We write down the numbers from 1 to 200. We then cross out all those which are divisible by 2 except 2 itself, then by 3 except 3 itself and so on. When we have deleted all the multiples of 13 except 13 itself we find that the only numbers remaining are the primes between 1 and 200. These are:

2, 3, 5, 7, *11*, *13*, *17*, *19*, 23, *29*, *31*, 37, *41*, *43*, 47, 53, *59*, *61*, 67, *71*, *73*, 79, 83, 89, 97, *101*, *103*, *107*, *109*, 113, 127, 131, *137*, *139*, *149*, *151*, 157, 163, 167, 173, *179*, *181*, *191*, *193*, *197*, *199*.

Were we to take these further we should notice that the prime numbers occur less frequently. This gives rise to the query as to whether there exists a greatest prime number. The proof that there is not is a beautiful piece of mathematical reasoning which is readily understood and appreciated by children at a certain stage. Suppose that the greatest prime number were P_n (i.e. the nth prime number). Consider $P_L = P_1 . P_2 . P_3 . . . . . P_n + 1$. P_L is divisible by all the known primes with remainder 1. Therefore P_L is either a prime or is divisible by a prime greater than P_n. Similarly, we can show that there exists a prime greater than P_L and so on. Hence there is no greatest prime number.

Another feature is that although the primes become less dense we still get pairs occurring near to each other. Between 1 and 200 there are no less than 13 pairs of primes which differ only by 2.

The next fact which strikes us is that every prime except 3 is either of the form $6n+1$ or $6n-1$ where n is an integer. Thus 7, 13, 19, 31, 37, etc., are of the form $6n+1$, and 5, 11, 17, 23, 29, etc., are of the form $6n-1$. The converse is not true. Numbers of the form $6n\pm 1$ are not necessarily primes. We need only mention 25 and 35 $(6.4+1, 6.6-1)$.

Attempts have been made to find a formula for constructing primes but so far none has been discovered. It was thought at one time that $2^{2^n}+1$ is prime for all n. This is true for $n = 1, 2, 3$ and 4, but $2^{2^5}+1$ or $2^{32}+1$ has since been factorized.

One final point of interest is the theorem that if p is prime and a is an integer not divisible by p, then $a^{p-1}-1$ is divisible by p. For example, if $a = 2, p = 5, a^{p-1}-1 = 15$ (which is divisible by 5), and if $a = 3, p = 7, a^{p-1}-1 = 728$ (which is divisible by 7).

POWERS OF NUMBERS

There is an abundance of problems which lead to geometrical progressions and many of these yield surprising results.

The story goes that Talleyrand, in return for services to the Austrian Emperor, was asked to name his own reward; nothing would be refused. Talleyrand pointed to the chessboard in front of him and requested that he might have a grain of wheat on the first square, two on the second, four on the third and so on for all the squares. It was a good jest, and a better one than at first appeared. Even the present annual world production of wheat would not satisfy such a demand!

If we conceive the situation in different terms and think of a penny instead of a grain of wheat we find that the total value will amount to $[1+2+4+8 \ldots +2^{63}]$ pence

or $$\left(\frac{2^{64}-1}{2-1}\right) \text{ pence}$$

or approximately £76,000,000,000,000,000.

Another surprising result occurs on folding paper. A piece of paper 0·01 in. thick, folded once becomes 0·01 × 2 in. thick. If it is folded 50 times, the thickness is $0{\cdot}01 \times 2^{50}$ in., or roughly 170,000,000 miles.

Geometrical progressions with a common ratio less than one also occur frequently in real life situations. Growth in plants and trees is generally exponential in form. A tree 10 ft. high is likely to increase its height by r per cent of its value each year. After n years the height is $10\left(1+\dfrac{r}{100}\right)^n$. On the other hand the value of a vehicle will decrease by, say, k per cent each year. Its value after n years is $£C\left(1-\dfrac{k}{100}\right)^n$ where $£C$ is the cost when new. Later in the chapter we consider in detail the graph of $y = 2^x$ and its application to logarithms and slide rules.

FRACTIONAL NUMBERS AND FAREY SERIES

Most children reach the secondary stage of their education with some ability to add and subtract simple fractions, but often they do not fully understand the process. A little revision is usually necessary but here, as in the other topics, a new approach is better than the medicine as before and patterns of an interesting sort abound. For example:

$$\frac{1}{1} - \frac{1}{3} = \frac{2}{3}$$

$$\frac{1}{3} - \frac{1}{5} = \frac{2}{3 \times 5}$$

$$\frac{1}{5} - \frac{1}{7} = \frac{2}{5 \times 7}$$

$$\frac{1}{7} - \frac{1}{9} = \frac{2}{7 \times 9}$$

Now, what is $\frac{1}{17} - \frac{1}{19}$?

and $\dfrac{1}{a} - \dfrac{1}{a-2}$?

Add the first four equations together. Can you state, without further working, the value of

$$\tfrac{1}{3}+\tfrac{1}{15}+\tfrac{1}{35}+\tfrac{1}{63}?$$

Again, in multiplication we have:

$$1\tfrac{1}{2}\times\tfrac{2}{3} = 1$$
$$2\tfrac{2}{3}\times\tfrac{3}{4} = 2$$
$$3\tfrac{3}{4}\times\tfrac{4}{5} = 3$$

Can you write down the next two without any working? What two fractions in this series have a product of 10?

It is also worth introducing the idea of a Farey series at this stage. The Farey series of order n is the series of all the rational fractions of the form $\dfrac{p}{q}$ in their lowest terms and arranged in order of magnitude such that $0 < \dfrac{p}{q} \leq 1; 0 < p \leq n; 0 < q \leq n$. For example the Farey series of order 3 (F_3) is

$$\tfrac{1}{3}, \ \tfrac{1}{2}, \ \tfrac{2}{3}, \ \tfrac{1}{1}$$

F_4 is $\ \tfrac{1}{4}, \ \tfrac{1}{3}, \ \tfrac{1}{2}, \ \tfrac{2}{3}, \ \tfrac{3}{4}, \ \tfrac{1}{1}$

F_5 is $\ \tfrac{1}{5}, \ \tfrac{1}{4}, \ \tfrac{1}{3}, \ \tfrac{2}{5}, \ \tfrac{1}{2}, \ \tfrac{3}{5}, \ \tfrac{2}{3}, \ \tfrac{3}{4}, \ \tfrac{4}{5}, \ 1$

Obviously a child will not be expected to write F_5 straight down. First of all we form all the possible fractions starting with denominator 1, 2, 3, 4 then 5. We then cross out all the repetitions and finally, by inspection (and mental arithmetic), we order the rest. Thus for F_5 we proceed

$$\tfrac{1}{1}, \ \tfrac{1}{2}, \ \cancel{\tfrac{2}{2}}, \ \tfrac{1}{3}, \ \tfrac{2}{3}, \ \cancel{\tfrac{3}{3}}, \ \tfrac{1}{4}, \ \cancel{\tfrac{2}{4}}, \ \tfrac{3}{4}, \ \cancel{\tfrac{4}{4}}, \ \tfrac{1}{5}, \ \tfrac{2}{5}, \ \tfrac{3}{5}, \ \tfrac{4}{5}, \ \cancel{\tfrac{5}{5}}$$

and finally rearrange as shown in the previous paragraph.

It is, of course, the ordering which requires most thought, but this can be done quite simply as well if we now introduce a graphical approach. We form a lattice of points at the integer-pairs satisfying the inequalities $y \leq x, \ y > 0, \ x \leq n$ where F_n is the series required. We then join every point to the origin. The

PATTERNS IN ARITHMETIC

gradients of these lines in order of magnitude from $\frac{1}{n}$ to $\frac{n}{n}$ will then give the Farey series F_n. Some lines overlap, but in such cases we only write down that gradient once. Thus, for F_4, see Fig. 112, i.e.
$$F_4 = \tfrac{1}{4},\ \tfrac{1}{3},\ \tfrac{1}{2},\ \tfrac{2}{3},\ \tfrac{3}{4},\ 1$$

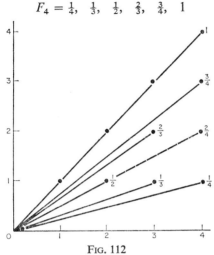

Fig. 112

Two points are worth noting. (i) In any Farey series, if $\frac{a}{b}, \frac{c}{d}$ are consecutive fractions, we find that $cb - ad = 1$. (ii) Given F_n we can form F_{n+1} by inserting between each pair of fractions $\frac{a}{b}, \frac{c}{d}$ in F_n a new fraction $\frac{a+c}{b+d}$ in its lowest terms. Now $\frac{a+c}{b+d}$ will, in some cases, have denominator greater than $n+1$, and these cases we exclude.

Thus, to form F_5 from F_4 we write out F_4 and insert fractions as shown, deleting those with denominator greater than 5.

$$F_4 \quad \tfrac{0}{1} \downarrow \tfrac{1}{4} \downarrow \tfrac{1}{3} \downarrow \tfrac{1}{2} \downarrow \tfrac{2}{3} \downarrow \tfrac{3}{4} \downarrow \tfrac{1}{1}$$
$$\tfrac{1}{5} \quad \cancel{\tfrac{2}{7}} \quad \tfrac{2}{5} \quad \tfrac{3}{5} \quad \cancel{\tfrac{5}{7}} \quad \tfrac{4}{5}$$
$$\therefore F_5 = \tfrac{1}{5},\ \tfrac{1}{4},\ \tfrac{1}{3},\ \tfrac{2}{5},\ \tfrac{1}{2},\ \tfrac{3}{5},\ \tfrac{2}{3},\ \tfrac{3}{4},\ \tfrac{4}{5},\ 1$$

THE FIBONACCI SEQUENCE

A Fibonacci sequence is one in which any term except the first is the sum of the two preceding terms. Thus, if we take the first term as unity (and assume that all preceding terms are zero) we have

1, 1, 2, 3, 5, 8, 13, 21, 34, 55, ...

This series arises in several interesting ways. It is named after Leonardo of Pisa, nicknamed Fibonacci (son of good nature), born 1175, and is reputed to have arisen out of his consideration of the following problem. Assume that rabbits live for ever and become mature one month after birth. One pair of mature rabbits produce a pair of young every month. Starting with one pair of immature rabbits how many mature pairs will there be at the end of each successive month thereafter? Denote a pair of immature rabbits by p and a pair of mature rabbits by P. The family tree is then represented diagrammatically as in Fig. 113.

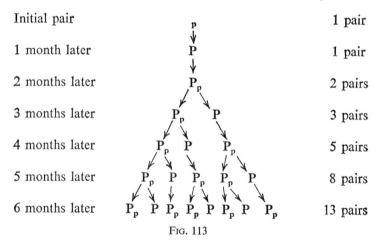

Fig. 113

The numbers of pairs of rabbits in successive months form a Fibonacci sequence.

An exactly similar pattern arises in the case of bees. A male bee

PATTERNS IN ARITHMETIC

or drone hatches from an unfertilized egg. Female workers or queens hatch from fertilized eggs. Thus, a male bee M has a mother F; a female bee F has a mother F and a father M. The family tree showing the ancestors of any particular drone is shown in Fig. 114. Once again we have the familiar sequence of numbers.

Professor Frank Land in his book *The Language of Mathematics* gives numerous other examples of the occurrence of this sequence in botany and musical scales.

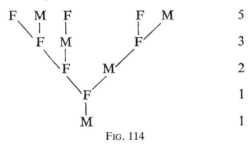

Fig. 114

We can obtain another picture of the pattern if we plot the values of the successive terms u_x as y values against x values 1, 2, 3, ... We obtain a curve which looks like an exponential curve and, in fact, after the first four terms, the curve is virtually indistinguishable from the exponential curve $y = 0.4472 \times 1.61803^x$. This approximation could be obtained by plotting a semilogarithmic graph of log (u_x) against integral x values. The result of course suggests that as x (integral) becomes large, the successive terms u_x, u_{x+1} approximate to the successive terms of a geometrical progression, or that the ratio $\dfrac{u_{x+1}}{u_x}$ tends to the limit 1·618 (correct to 3 decimal places).

It is a useful exercise, and one which illustrates the idea of a limit, to investigate the ratios of successive terms of the Fibonacci sequence. As we see below, these quickly converge upon the steady value 1·618.

$\frac{1}{1} = 1·000$ $\quad$ $\frac{5}{3} = 1·667$ $\quad$ $\frac{21}{13} = 1·615$ $\quad$ $\frac{55}{34} = 1·6176$
$\frac{2}{1} = 2·000$ $\quad$ $\frac{8}{5} = 1·6$ $\quad\quad$ $\frac{34}{21} = 1·619$ $\quad$ $\frac{89}{55} = 1·6182$
$\frac{3}{2} = 1·500$ $\quad$ $\frac{13}{8} = 1·625$

and so on. In fact, if we consider sequences in which any term except the first *two* is the sum of the two preceding terms thus:

$$a, \ b, \ a+b, \ a+2b, \ 2a+3b, \ \ldots$$

we obtain a class of sequences closely related to the Fibonacci sequence. These all possess the property that the limiting ratio of successive terms is 1·6182.

Arithmetical and geometrical progressions have general terms which are easily formulated, but the "disadvantage" of the Fibonacci sequence, and probably one reason why it rarely appears in school mathematics syllabuses, is that the formula for the nth term u_n is

$$\frac{1}{\sqrt{5}}\left(\frac{1+\sqrt{5}}{2}\right)^n - \frac{1}{\sqrt{5}}\left(\frac{1-\sqrt{5}}{2}\right)^n$$

hardly a result which one could expect children to discover for themselves! Nevertheless it is a good exercise to evaluate the expression for small integral values of n.

It does, however, help us to see why the curve of u_n values approximated closely to $y = 0.4472 \times 1.618^n$. For as $n \to \infty$,

$$\left(\frac{1-\sqrt{5}}{2}\right)^n \to 0 \quad \text{and} \quad u_n \to \frac{1}{\sqrt{5}}\left(\frac{1+\sqrt{5}}{2}\right)^n$$

i.e. $\qquad u_n \sim 0.4472 \times 1.618^n$

THE GOLDEN SECTION

The numbers $t = \dfrac{1+\sqrt{5}}{2}$, $t' = \dfrac{1-\sqrt{5}}{2}$ from which $tt' = -1$, $t+t' = 1$ are clearly the roots of the quadratic $t^2 - t - 1 = 0$. Writing $t = l/w$ this equation gives:

$$wl + w^2 = l^2$$

or
$$\frac{w}{l} = \frac{l}{w+l}$$

Consider now a rectangle, length l units and width w units, in which $\dfrac{\text{width}}{\text{length}} = \dfrac{\text{length}}{\text{width}+\text{length}}$ and we see that the ratio of the length to the width $\frac{l}{w}$ is t or $\frac{1+\sqrt{5}}{2}$. A rectangle in which these proportions obtain is called a golden rectangle and the ratio t, $\frac{1+\sqrt{5}}{2}$ or 1·618 is called the golden number.

It is an interesting experiment to draw several rectangles on the blackboard, including one in which the ratio of adjacent sides is roughly 1·62 and then to take a vote on the most pleasantly proportioned one. In the writer's experience the golden rectangle usually tops the poll.

The Greeks were well aware of the beautiful proportions of this rectangle and many of their buildings illustrate it.

An extraordinary quality of this rectangle is that it may be cut into two parts, one a square and the other a golden rectangle. For if the original rectangle has adjacent sides of length t and 1 unit and we cut off a square of side 1 unit the remaining rectangle has sides $t-1$ and 1 unit. But since $t^2-t-1 = 0$ we have $\frac{1}{t-1} = \frac{t}{1}$, therefore the remaining rectangle is golden. This process can be continued indefinitely until the whole rectangle has been sectioned into a series of squares whose sides are in geometrical progression with common ratio $\frac{1}{t}$, together with one infinitesimally small golden rectangle.

We shall not deal with the golden spiral in this chapter. A treatment of this may be found in *Introduction to Geometry* by H. S. M. Coxeter. However, at an elementary stage, if we construct circular quadrants in all the squares mentioned above, we obtain a close approximation to the golden spiral and this curve also has interesting properties. From Fig. 115 we can derive several interesting series involving the golden number t. For example, the area of the whole rectangle is equal to the sum of an infinite number of squares:

$$t = 1+\frac{1}{t^2}+\frac{1}{t^4}+\frac{1}{t^6}+\frac{1}{t^8}\ldots$$

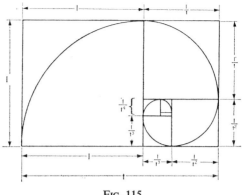

Fig. 115

a result which can be obtained independently from the relation

$$t - 1 = \frac{1}{t},$$

$$\therefore t = \left(1 - \frac{1}{t^2}\right)^{-1} = 1 + \frac{1}{t^2} + \frac{1}{t^4} + \frac{1}{t^6} \ldots$$

Incidentally, the relation $t - \frac{1}{t} = 1$ enables us to express t in the form of a continued fraction. Thus, if $t = 1 + \frac{1}{t}$ resubstituting we have

$$t = 1 + \frac{1}{1 + 1/t}$$

$$= 1 + \frac{1}{1 + \frac{1}{1 + 1/t}}$$

i.e.
$$t = 1 + \frac{1}{1+} \frac{1}{1+} \frac{1}{1+} \frac{1}{1+} \ldots$$

Further, the length of the "spiral" shown in the diagram is

$$\frac{\pi}{2}\left[1+\frac{1}{t}+\frac{1}{t^2}+\frac{1}{t^3}\cdots\right]$$

or $\quad\dfrac{\pi}{2}\left[\dfrac{1}{1-1/t}\right]\quad$ or $\quad\dfrac{\pi}{2}\cdot\dfrac{t}{t-1}\quad$ or $\quad\dfrac{\pi t^2}{2}$

Thus $\qquad t^2 = 1+\dfrac{1}{t}+\dfrac{1}{t^2}+\dfrac{1}{t^3}\cdots$

Another way in which the number t arises may be illustrated with reference to a regular pentagon of unit side (Fig. 116).

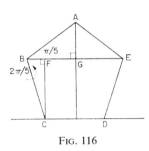

Fig. 116

For $\qquad BE = 2\cos\dfrac{\pi}{5}$

but $\qquad BF = BG - FG$

$\therefore\ \cos\dfrac{2\pi}{5} = \cos\dfrac{\pi}{5} - \dfrac{1}{2}$

$\therefore\ 2\cos^2\dfrac{\pi}{5} - 1 = \cos\dfrac{\pi}{5} - \dfrac{1}{2}$

$\therefore\ 4\cos^2\dfrac{\pi}{5} - 2\cos\dfrac{\pi}{5} - 1 = 0$

Put $T = 2 \cos \frac{\pi}{5}$ and we obtain once more the equation

$$T^2 - T - 1 = 0$$

i.e. T is equal to the golden number.

∴ BE is of length t, the golden number.

Further, the point of intersection of any two diagonals of the regular pentagon is the point of golden section of either diagonal.

Finally, consider the sequence of numbers which arises if we evaluate successive powers of t. Here is a way in which practice in the manipulation of surds may be used to discover further mathematical patterns.

$$t = \frac{\sqrt{5}+1}{2} \qquad t^4 = \frac{3\sqrt{5}+7}{2}$$

$$t^2 = \frac{\sqrt{5}+3}{2} \qquad t^5 = \frac{5\sqrt{5}+11}{2}$$

$$t^3 = \frac{2\sqrt{5}+4}{2} \qquad t^6 = \frac{8\sqrt{5}+18}{2}$$

The coefficients of $\sqrt{5}$ in successive powers of t form the Fibonacci sequence and the constant terms in the numerators 1, 3, 4, 7, 11, 18, ... form another Fibonacci sequence from the third term onwards.

INDICES AND LOGARITHMS

Consider the exponential function 2^x. Expressed in another way this is a mapping of the set $\{x \mid x \in R\}$ on to the set $\{2^x \mid x \in R\}$. The set $\{2^x \mid x \text{ an integer}\}$ is a group under multiplication. The elements of this group, for which x is a positive integer, provide an example of a geometrical progression of common ratio 2, and the graph of the function provides a picture of a growth law, e.g. the way in which the thickness of a paper increases with folding. Further, we can take the function 2^x as a starting point for the study of logarithms.

PATTERNS IN ARITHMETIC

There is clearly a pattern in Table 49 showing values of x and 2^x.

TABLE 49

Power (x)	Number 2^x	Denary	Binary
5	$2 \times 2 \times 2 \times 2 \times 2$	32	100,000
4	$2 \times 2 \times 2 \times 2$	16	10,000
3	$2 \times 2 \times 2$	8	1,000
2	2×2	4	100
1	2	2	10
0	1	1	1
−1	$\frac{1}{2}$	0·5	0·1
−2	$\frac{1}{2} \times \frac{1}{2}$	0·25	0·01
−3	$\frac{1}{2} \times \frac{1}{2} \times \frac{1}{2}$	0·125	0·001
−4	$\frac{1}{2} \times \frac{1}{2} \times \frac{1}{2} \times \frac{1}{2}$	0·0625	0·0001
−5	$\frac{1}{2} \times \frac{1}{2} \times \frac{1}{2} \times \frac{1}{2} \times \frac{1}{2}$	0·03125	0·00001

One way of illustrating this pattern is by means of a graph (Fig. 117).

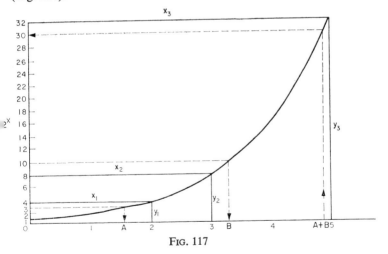

FIG. 117

Now consider the product $2^2 \times 2^3 = 4 \times 8 = 32 = 2^5$. We notice that $\quad 4 \times 8 = 32$ (numbers)

and $\quad 2+3 = 5$ (indices or logarithms)

or, from the diagram, $y_1 \times y_2 = y_3$
and $x_1 + x_2 = x_3$

Similarly, for any three points on the graph, if

$$y_p + y_q = y_r \quad \text{then} \quad x_p + x_q = x_r$$

It follows that we can use this graph to perform multiplications. If we wish to *multiply* a number y_1 by a number y_2, we simply add x_1 and x_2 and read off the value of $y_1 \times y_2$ as the ordinate corresponding to $x_1 + x_2$. To multiply the number 3 by the number 10, for example, find the x value A corresponding to 3 on the vertical scale, the x value B corresponding to 10 and obtain the ordinate corresponding to $x = A + B$. In this case $2^A = 3, 2^B = 10$; A and B are the logarithms to the base 2 of 3 and 10 respectively.

THE BINARY SLIDE RULE

A quicker way of performing these multiplications is to construct a simple slide-rule. Using the same graph, we now renumber the x scale as shown in Fig. 118. The scale is marked off on two pieces of cardboard, wood or plastic which are then fastened together in such a way that one can slide freely on the other (Fig. 119). (This is a good opportunity for co-operation

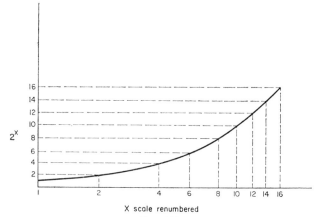

X scale renumbered

FIG. 118

between the Mathematics and Craft Departments in the first or second year of the secondary school course.) To multiply 3 by 4 we simply place the 1 on the bottom scale under the 3 on the top scale and read off the answer above the 4 on the bottom scale.

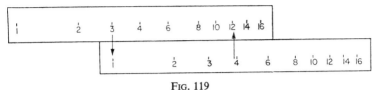

Fig. 119

The same procedure could be performed in the scale of 10, but in this case the need to evaluate $10^{\frac{1}{2}}$ and $10^{\frac{1}{4}}$ arises long before the pupils are able to extract such roots for themselves. In any case this approach ties in well with binary arithmetic and other scales of notation.

THE BINARY SCALE AND OTHER SCALES

In the denary scale the number 111 means $10^2 + 10^1 + 10^0$. The place values are successive powers of 10. This mode of counting and expressing numbers probably derives from the fact that we have ten fingers. In a world of intelligent creatures possessing two digits it is possible that a system of place values based on powers of 2 might evolve. Such a scale is called *binary* and on this scale the number 111 means $2^2 + 2^1 + 2^0$. That is to say 111 on the binary scale is the same number as 7 on the denary scale. Alternatively the denary number $111 = 64 + 32 + 8 + 4 + 2 + 1 = 2^6 + 2^5 + 2^3 + 2^2 + 2^1 + 2^0 +$ the binary number 1101111.

To change decimal fractions into *binary* fractions we proceed as in the following examples:

$$0.75 = \tfrac{1}{2} + \tfrac{1}{4} = 2^{-1} + 2^{-2} = 0.11 \text{ (binary scale)}$$
$$0.7 = \tfrac{7}{10} = \tfrac{5}{10} + \tfrac{2}{10} = \tfrac{5}{10} + \tfrac{16}{80}$$
$$= \tfrac{5}{10} + \tfrac{10}{80} + \tfrac{5}{80} + \tfrac{10}{1280} + \tfrac{1}{256} + \cdots$$
$$= 2^{-1} + 2^{-3} + 2^{-4} + 2^{-7} + 2^{-8} \cdots$$
$$= 0.10110011 \ldots \text{ in the binary scale.}$$

In the second case it is quicker to divide 7 by 10 in the binary scale:

```
              0·10110011 ...
        ┌─────────────────
   1010 │ 111·0
        │ 101 0
        │ ─────
        │  10 000
        │   1 010
        │   ─────
        │    1100
        │    1010
        │    ────
        │     10000
        │      1010
        │      ────
        │       1100
        │       1010
        │       ────
        │         10, etc.
```

Since numbers in the binary scale are expressed only in terms of the digits 1 and 0, they may be represented by the presence or absence of a current flowing in a circuit. It is for this reason that electronic computers work with binary numbers.

A binary adder, whose construction is well within the capacity of the average sixth-form pupil, was devised by Dr H. M. Cundy and described in the *Mathematical Gazette*, Vol. XLII, No. 342, December 1958. There is a similar description in Cundy and Rollett's excellent book *Mathematical Models*, 2nd ed. (Oxford).

In a simple adder, capable of a five-digit answer, the numbers to be added are represented by switches on (1) or off (0), and the result is given by five light bulbs (Fig. 120).

A light on represents 1.

Binary	Denary
1001+	9+
1101	13
10110	22

The binary scale has been applied to signal systems.† As an alternative to the Morse code, which requires long pulses and

† See Chapter 2, *Some Lessons in Mathematics*, ed. T. J. Fletcher.

PATTERNS IN ARITHMETIC

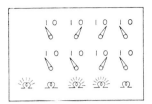

Fig. 120

short pulses, signal systems have been designed assigning binary numbers to the various letters of the alphabet so that each letter is transmitted as a different single combination of pulses (unit digits) and pauses (zero digits). The simplest binary numbers correspond to the letters which are used most frequently.

In his book *Fundamental Concepts of Mathematics* Professor Goodstein discusses the classical problem where twelve objects are given, eleven of which are equal in weight and one lighter or heavier than the rest. This latter object has to be identified by three weighings. The solution involves labelling the weights with three digit numbers expressed in the scale of three.

A successful player in the game of Nim must also be able to use the binary scale. In this game two players draw articles from three piles, each player in turn taking as many as he likes from any one pile. The player drawing the last article wins. In order to win, a player must, after each draw, leave such numbers of articles in each pile that, when expressed in the binary scale, the sum of the corresponding digits is either 0 or 2. For example if the piles contain

$$\begin{array}{rll} 8 & \text{i.e.} & 1000 \\ 4 & \text{i.e.} & 100 \\ 3 & \text{i.e.} & 11 \end{array}$$

the winner will leave

$$\begin{array}{rll} 7 & \text{i.e.} & 111 \\ 4 & \text{i.e.} & 100 \\ 3 & \text{i.e.} & 11 \end{array}$$

No matter what the other player now takes he must leave one column odd. The winner now makes all columns even again and whatever his opponent does he will eventually leave two piles with one object in each. He must, therefore, take the last object and win.

Practice in scales of notation need not be confined to the binary scale. In any case children already use other scales. The addition of pence involves a scale of 12, and with hours, minutes and seconds we have a scale of 60. Indeed, we need not use familiar symbols at all. Consider an ordered system of numbers:

$$| \quad + \quad I \quad \mp \quad X \quad \ni \quad +| \quad ++ \quad +I \quad \text{etc.}$$

corresponding to

$$0 \quad 1 \quad 2 \quad 3 \quad 4 \quad 5 \quad 6 \quad 7 \quad 8 \quad \text{etc.}$$

This is essentially a scale of six, and, for example:

$$11 = 6+5 = +\ni$$
$$94 = 2.6^2 + 3.6^1 + 4.6^0 = I\mp X$$
$$\ni + \mp = 5.6^2 + 1.6^1 + 3.6^0 = 189$$
$$\ni \mp + \text{minus } X | I = +I\ni$$

In this chapter we have tried to show a few of the beautiful and instructive patterns to be found in numbers. The subject is as vast as it is fascinating and no book can reveal more than a glimpse of it. The problem of computation alone, the methods of numerical analysis and the development of computers is a huge expanding subject in its own right and it is already revolutionizing our science and technology and, indeed, our attitude to mathematics itself. We have seen the drudgery removed from many industrial processes and household chores. It is time we employed our own artifacts to do the chores of school arithmetic and freed ourselves to contemplate the beauties of its structures.

Useful Reference Books

ON COMPUTING AND NUMERICAL ANALYSIS

BOWDEN, B. V. (ed.), *Faster than Thought*, Pitman.
IRWIN, W. C., *Digital Computer Principles*, Van Nostrand.
LOVIS, F. B., *Computers I and II*, E. Arnold.
MOAKES, A. J., *Numerical Analysis*, Macmillan.
THOMAS and THOMAS, *Mathematics by Calculating Machine*, Cassell.
WOOLDRIDGE, R., *An Introduction to Computing*, Oxford University Press.

On Patterns, Problems and Puzzles

ANDREWS, W. S., *Magic Squares and Cubes*, Dover.
FIELKAR, D. S., Now do the following fifty examples, *Mathematics Teaching* Spring 1963.
HOGBEN, L., *Mathematics for the Million*, Allen & Unwin.
HOOPER, A., *Makers of Mathematics*, Faber & Faber.
KASNER, E. and NEWMAN, I., *Mathematics and the Imagination*, Bell.
KRAITCHIK, M., *Mathematical Recreations*, Allen & Unwin.
LAND, F., *The Language of Mathematics*, John Murray.
MERRILL, H. A., *Mathematical Excursions*, Constable.
NORTHROP, E. P., *Riddles in Mathematics*, English Universities Press.
REICHMANN, W. J., *The Fascination of Numbers*, Methuen.
REID, C., *From Zero to Infinity*, Routledge & Kegan Paul.
ROUSE BALL, W. W., *Mathematical Recreations and Essays*, Macmillan.
STEINHAUS, H., *Mathematical Snapshots*, Oxford.

APPENDIX

Exercises on Chapters 1–7

EXERCISES ON SETS

1. The days on which John has mathematics form the set $M = \{$Monday, Tuesday, Thursday, Friday$\}$ and the days on which he has physics form the set $P = \{$Tuesday, Wednesday, Friday$\}$. Taking the days of the week as the universal set $\mathscr{E}$ list the following sets:

(a) $M \cup P$, (b) $M \cap P$, (c) M', (d) P', (e) $\mathscr{E} \cap P$, (f) $\mathscr{E} \cup M$, (g) $M' \cap P'$, (h) $M' \cup P'$, (i) $(M \cup P)'$, (j) $(M \cap P)'$.

Which of these sets are equal?

2. Four house captains are H or $\{$Alfred, David, Eric, Harold$\}$; the tennis six is S or $\{$Brian, Colin, David, Fred, George, Ian$\}$; the rowing eight is R or $\{$Alfred, Brian, Colin, David, Eric, Ian, Harold, John$\}$. Taking the ten boys as a universal set $\mathscr{E}$, list, by initials, the following sets:

(a) $H \cap S$, (b) $S \cap R$, (c) $H \cap R$, (d) $H \cap S \cap R$, (e) $H \cup S \cup R$, (f) $H \cap (S \cup R)$, (g) $(H \cap S) \cup (H \cap R)$, (h) $H \cup (S \cap R)$, (i) $(H \cup S) \cap (H \cup R)$.

Which of these sets are equal?

3. $P = \{a, b, c, d\}$, $Q = \{a, b, c\}$, $R = \{a, b\}$. $A \triangle B$ is the set of those elements in A or in B but not in both. Write down the sets $P \cap (Q \triangle R)$ and $(P \cap Q) \triangle (P \cap R)$. Are they equal? What about $P \triangle (Q \cap R)$ and $(P \triangle Q) \cap (P \triangle R)$?

4. For the universe $\mathscr{E} = \{x \mid x \text{ is a positive integer}\}$ list the subsets of $\mathscr{E}$ which satisfy the following:

(a) $x+4 = 6$, (b) $x+4 = 2$, (c) $2x+8 = 2(x+4)$,

(d) $2x-3 = 0$, (e) $x+2 > 3$, (f) $x+6 > 2$, (g) $x+3 < 4$.

5. If $\mathscr{E}$ is the set of integers and $x \in \mathscr{E}$, list the following sets:

(a) $\{x \mid x \geqq 2\} \cup \{x \mid x < 2\}$, (b) $\{x \mid x \geqq 2\} \cap \{x \mid x \leqq 2\}$,

(c) $\{x \mid 1 < x < 5\} \cap \{x \mid 0 < x < 3\}$,

(d) $\{x \mid x \leqq 4\} \cap \{x \mid 2 \leqq x \leqq 5\} \cap \{x \mid x \geqq 3\}$,

(e) $\{x \mid x < 4\} \cap \{x \mid 2 \leqq x \leqq 5\} \cap \{x \mid x \geqq 3\}$,

(f) $\{x \mid x < 4\} \cap \{x \mid 2 < x < 5\} \cap \{x \mid x > 3\}$,

(g) $\{x \mid x > 4\} \cup \{x \mid 2 < x < 5\} \cup \{x \mid x < 3\}$.

6. If $\mathscr{E} = \{1, 2, 3, 4\}$ mark in the set of lattice points of ordered pairs of the product set $\mathscr{E} \times \mathscr{E}$. For this inverse, and using a separate diagram in each case, ring those ordered number pairs which comprise the following sets:

(a) $\{(x,y) \mid y = x\}$, (b) $\{(x,y) \mid y = x+1\}$,

(c) $\{(x,y) \mid y \geqq x\}$, (d) $\{(x,y) \mid y < x\}$,

(e) $\{(x,y) \mid x+y = 6\} \cap \{(x,y) \mid 2x-3y = 2\}$,

(f) $\{(x,y) \mid (x-2)(y-4) = 0\}$,

(g) $\{(x,y) \mid x > 1\} \cap \{(x,y) \mid x+y \leqq 4\} \cap \{(x,y) \mid y > 1\}$,

(h) $\{(x,y) \mid x^2+y^2 < 25\}$.

7. For all real values of x and y, draw the graphs of the following sets:

(a) $\{(x,y) \mid y = x-1\}$, (b) $\{(x,y) \mid x+y = 4\}$,

(c) $\{(x,y) \mid x = 3\}$, (d) $\{(x,y) \mid y = 2\}$, (e) $\{(x,y) \mid xy = 4\}$

(f) $\{(x,y) \mid y < x^2 \text{ and } -1 < x < 1\}$,

(g) $\{(x,y) \mid x+y < 5\} \cap \{(x,y) \mid 5y+2x > 10\} \cap$
    $\{(x,y) \mid 2x+y > 6\}$.

8. In a group of 50 students, 15 play tennis, 20 play cricket and 20 do athletics. 3 play tennis and cricket, 6 play cricket and do athletics and 5 play tennis as well as doing athletics. 7 take no part in games. Now many play cricket, tennis and do athletics?

9. Of 100 vehicles taking Ministry of Transport tests, 60 passed. Amongst the remainder, faults in brakes, lights and steering occurred as follows:

Brakes only	12
Brakes and steering	5
Brakes, steering and lights	3
Brakes and lights	8
Steering and lights only	2

Equal numbers of cars having one fault only, failed because of steering or lights.

How many cars had faulty lights?
How many cars had only one fault?

10. If $f_1(z) = z$, $f_2(z) = -z$, $f_3(z) = \dfrac{1}{z}$, $f_4(z) = -\dfrac{1}{z}$, state, as single functions, the following:

(a) $f_2 f_3(z)$, (b) $f_2 f_4(z)$, (c) $f_3 f_3(z)$, (d) $f_1 f_4(z)$, (e) $f_4 f_4(z)$, (f) $f_2 f_3 f_4(z)$.

EXERCISES ON LOGIC AND BOOLEAN ALGEBRA

1. In which of the following arguments does the conclusion necessarily follow from the premises? If it does write T (true), if it does not, F (false). Venn diagrams may be used where helpful.

(a) Some gems are expensive.
All diamonds are expensive.
Therefore some gems are diamonds.

(b) Some rare gases are inert.
Argon is a rare gas.
Therefore Argon is an inert gas.

(c) All sculptors are artists.
Some Frenchmen are sculptors.
Therefore some Frenchmen are artists.

(d) All sculptors are artists.
Some Frenchmen are sculptors.
Therefore some French artists are not sculptors.

(e) Some teachers teach mathematics or physics or chemistry.
Some mathematics teachers can teach physics.
Some physics teachers can teach chemistry.
Therefore some mathematics teachers can teach chemistry.

(f) Animals are either wild or tame.
Animals are either large or small.
Therefore some tame animals are large.

(g) Animals are either wild or tame.
Animals are either large or small.
Some tame animals are large.
Therefore some wild animals are small.

(h) Mathematical problems are either algebraic or geometrical.
Mathematical problems are either easy or difficult.
Some geometrical problems are easy and some are difficult.
Therefore some algebraic problems are difficult.

(i) $x < 1$
$y < 1,000$
Therefore $x < y$

(j) $x < 2$
$x > 7$
Therefore $7 < 2$

(k) $y < x$
$y^2 > x^2$
Therefore $x < 0$

EXERCISES ON CHAPTERS 1–7

(l) Some cyclic quadrilaterals (C) are trapezia (T).
Some rhombuses (R) are cyclic quadrilaterals.
Therefore some trapezia are rhombuses.
Is the conclusion (a) valid (b) true?
Describe the figures in $C \cap T \cap R$

2. Given the following premises state the main conclusion:

(a) In our village all the white cows are branded J.
Cows branded J belong to Farmer Jones.
Only black cows wear bells.

(b) No intelligent people are bad-tempered.
Some intelligent people are red-haired.

(c) Some old things are valuable.
Valuable things are beautiful.

3. Simplify the following set of rules, all of which must be obeyed:

Boys may wear no tie and/or no white shirt.
Boys may wear no tie and/or a white shirt.
Boys may wear a tie and/or a white shirt.

4. If a represents the statement "I like tennis", and b represents the statement "I am a good cricketer", write out the meaning of the following:

(i) $a \wedge b$. (ii) $a \vee b$. (iii) $a \wedge \sim b$. (iv) $\sim a \vee b$. (v) $\sim a \wedge \sim b$.
(vi) $a \rightarrow b$.

5. Write out a truth table for each of the following:

(i) $p \vee \sim p$. (ii) $p \vee \sim q$. (iii) $p \wedge \sim q$. (iv) $\sim p \wedge q$.
(v) $(p \wedge \sim q) \vee (\sim p \wedge q)$.

6. Give the Boolean function for each of the following switching circuits (Figs. 121–9) and simplify the function (and hence the circuit) where possible:

(a)

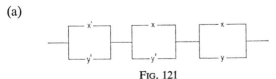

Fig. 121

(b)

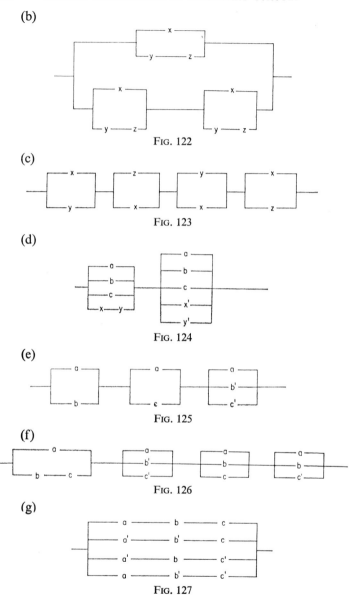

Fig. 122

(c)

Fig. 123

(d)

Fig. 124

(e)

Fig. 125

(f)

Fig. 126

(g)

Fig. 127

(h)

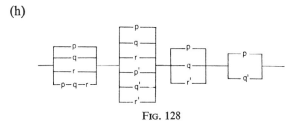

Fig. 128

(i)

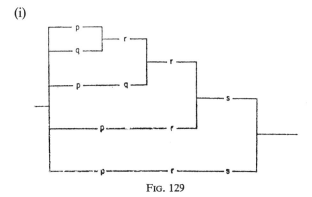

Fig. 129

7. Write out a closure table for 6(g) and state to what use this circuit may be put.

8. A machine contains three fuses p, q, r. It is desired to arrange them so that if p blows the machine stops, but if p does not blow then the machine only stops when both q and r have blown. Design the required fuse circuit.

9. A committee of five vote by pressing switch buttons a, b, c, d and e. The chairman, who controls switch a, does not vote unless required to give a casting vote. Otherwise the decision is by a majority vote of the other four numbers. Design a switching circuit by means of which a buzzer sounds whenever a resolution is carried.

EXERCISES ON GROUPS

(The text of Chapter 3 itself contains several suggested exercises for the reader.)

1. Show that $\{1, e^{i\pi/3}, e^{2i\pi/3}, e^{3i\pi/3}, e^{4i\pi/3}, e^{5i\pi/3}\}$ forms, under multiplication, a cyclic group of order 6. Write out its multiplication table. Is it a commutative group?

2. Form the addition table of the group of integers $\{0, 1, 2, 3, 4, 5\}$ mod 6, and show that this group is isomorphic with the group in Question 1.

3. Generalize the results of Question 1 and Question 2.

4. Form the multiplication table of the permutation group of order 2. Show that all groups of order 2 are isomorphic.

5. For the six functions $f_1(x) = x$, $f_2(x) = \dfrac{1}{1-x}$, $f_3(x) = \dfrac{x-1}{x}$, $f_4(x) = \dfrac{1}{x}$, $f_5(x) = 1-x$, $f_6(x) = \dfrac{x}{x-1}$, we define the law of composition as the substitution of one function in another. For example:

$$f_2 f_3 = f_2\{f_3(x)\} = \frac{1}{1 - \dfrac{x-1}{x}} = x = f_1(x) \quad \text{or} \quad f_1$$

$$f_3 f_4 = f_3\{f_4(x)\} = \frac{\dfrac{1}{x} - 1}{\dfrac{1}{x}} = 1 - x = f_5(x) \quad \text{or} \quad f_5$$

Obtain all such "products" and show that $\{f_1 f_2 f_3 f_4 f_5 f_6\}$ form a non-commutative group.

> The functions $f_1 \ldots f_6$ represent the six possible values of the cross-ratio of four points. Written as $f_1(z) = z$ where $z = x + iy$, they each have the property of mapping the complex plane into itself.

EXERCISES ON CHAPTERS 1–7

6. Show that the group of integers mod 6 in Question 2 has subgroups of order 2 and order 3 and list them.

7. Show that the group of rotations of the equilateral triangle $\{1, \omega, \omega^2\}$ is isomorphic with a subgroup of the permutation group S_3 and list this subgroup.

8. If $f_1(z) = z$, $f_2(z) = -z$, $f_3(z) = \dfrac{1}{z}$, $f_4(z) = -\dfrac{1}{z}$, show that the substitution group $\{f_1 f_2 f_3 f_4\}$ has three subgroups of order 2. State which subgroups of $\{1, 3, 5, 7\}$ mod 8 under multiplication are respectively isomorphic with these three groups.

9. Show that the matrices
$$\left\{ \begin{pmatrix} 1 & 0 \\ 0 & 1 \end{pmatrix}, \begin{pmatrix} \tfrac{1}{2} & -\sqrt{3}/2 \\ \sqrt{3}/2 & \tfrac{1}{2} \end{pmatrix}, \begin{pmatrix} -\tfrac{1}{2} & -\sqrt{3}/2 \\ \sqrt{3}/2 & -\tfrac{1}{2} \end{pmatrix}, \begin{pmatrix} -1 & 0 \\ 0 & -1 \end{pmatrix}, \begin{pmatrix} -\tfrac{1}{2} & \sqrt{3}/2 \\ -\sqrt{3}/2 & -\tfrac{1}{2} \end{pmatrix}, \begin{pmatrix} \tfrac{1}{2} & \sqrt{3}/2 \\ -\sqrt{3}/2 & \tfrac{1}{2} \end{pmatrix} \right\}$$
form a group under matrix multiplication and show that this group is isomorphic with the groups in Questions 1 and 2.

EXERCISES ON MATRICES

1. P is the point $(1, 1)$ and Q is the point $(2, 2)$. Transform P and Q by means of the following matrices:

(a) $\begin{pmatrix} 1 & 0 \\ 0 & 1 \end{pmatrix}$ (b) $\begin{pmatrix} 1 & 0 \\ 0 & -1 \end{pmatrix}$ (c) $\begin{pmatrix} -1 & 0 \\ 0 & 1 \end{pmatrix}$

(d) $\begin{pmatrix} 1/\sqrt{2} & -1/\sqrt{2} \\ 1/\sqrt{2} & 1/\sqrt{2} \end{pmatrix}$ (e) $\begin{pmatrix} 1 & -1 \\ 1 & 1 \end{pmatrix}$ (f) $\begin{pmatrix} 2 & 0 \\ 0 & 2 \end{pmatrix}$

(g) $\begin{pmatrix} 2 & 1 \\ 1 & 2 \end{pmatrix}$

Describe the transformation of the segment PQ in each case in terms of translation, reflection, rotation and enlargement.

2. If $\begin{array}{lll} x_3 = x_2 + y_2 & x_2 = x_1 + 2y_1 & x_1 = 2x_0 - y_0 \\ y_3 = x_2 - y_2 & y_2 = 2x_1 - y_1 & y_1 = x_0 + 2y_0 \end{array}$

express x_3, y_3 in terms of x_0 and $2y_0$

3. Construct the matrix which, operating upon the segment OP where P is the point $(1, 1)$,

   (a) doubles its length,
   (b) triples its length,
   (c) doubles its length and rotates it anti-clockwise through $\pi/3$ radians,
   (d) gives its reflection in the X-axis,
   (e) gives its reflection in the Y-axis,
   (f) translates it into a parallel equal segment, joining the points $(3, 4)$ $(4, 5)$.

4. What is the effect of the matrix $\begin{pmatrix} 1 & 1 \\ 1 & 1 \end{pmatrix}$ upon the square, vertices $(0, 0)$ $(1, 0)$ $(1, 1)$ $(0, 1)$?

5. Apply the transformation $\begin{pmatrix} X \\ Y \end{pmatrix} = \begin{pmatrix} 1 & 0 \\ 0 & k \end{pmatrix} \cdot \begin{pmatrix} x \\ y \end{pmatrix} (0 < k < 1)$ to the unit circle $x^2 + y^2 = 1$, and show that the resulting figure is an ellipse with a major semi-axis one unit long. For what value of k has

   (a) the ellipse half the area of the circle,
   (b) the ellipse an eccentricity of $\tfrac{1}{2}$?

6. $$A = \begin{pmatrix} 1 & 2 & 3 \\ 3 & 1 & 2 \end{pmatrix} \quad B = \begin{pmatrix} 2 & 5 \\ 3 & 0 \\ 4 & 1 \end{pmatrix}$$

Form the matrices (a) $A.B$ (b) $B.A$.

7. If $A = \begin{pmatrix} 2 & 3 \\ 1 & 4 \end{pmatrix}$ $B = \begin{pmatrix} 1 & 4 \\ 2 & 3 \end{pmatrix}$ form the matrices

(a) $A.B$, (b) $B.A$, (c) $A.\tilde{A}$, (d) $B.\tilde{B}$ (e) $\tilde{A}.\tilde{B}$ (f) $\widetilde{B.A}$, and evaluate (g) $|A|.|B|$, (h) $|A.B|$.

8. $$A = \begin{pmatrix} 1 & 0 & 0 \\ 0 & 2 & 0 \\ 0 & 0 & 3 \end{pmatrix} \quad B = \begin{pmatrix} 3 & 0 & 0 \\ 0 & 1 & 0 \\ 0 & 0 & 2 \end{pmatrix}$$

Form the matrices

(a) $A.B$, (b) $B.A$, (c) $\tilde{A}.\tilde{B}$, (d) $\widetilde{B.A}$, and evaluate (e) $|A|.|B|$, (f) $|A.B|$.

9. $$A = \begin{pmatrix} 1 & 1 \\ 2 & 2 \end{pmatrix} \quad B = \begin{pmatrix} 1 & -1 \\ -1 & 1 \end{pmatrix}.$$

Form the matrices (a) $A.B$, (b) $B.A$. Show that in each case the rule $|P|.|Q| = |P.Q|$ still holds.

10. Write $\begin{pmatrix} 2 & 4 & 6 \\ 8 & 10 & 12 \\ 14 & 16 & 18 \end{pmatrix}$ as the sum of two matrices, one symmetric and the other anti-symmetric.

11. Find the inverse of the matrix $\begin{pmatrix} 4 & -2 \\ 3 & 1 \end{pmatrix}$. Use this inverse matrix to solve the following simultaneous equations:

(a) $4x - 2y = p$ (b) $2x - y = 6$ (c) $y = 2x$
$3x + y = q$ $3x + y = 14$ $y = 5 - 3x$
(d) $4x - 2y = 5 = 6x + 2y$.

12. Solve the equations

$$x + z = 3$$
$$y + z = 2$$
$$x + 3y + 2z = 7$$

by the inverse matrix method.

13. The cost of manufacturing three types of motor-car is given by the following matrix:

	Labour (hours)	Materials (units)	Subcontracted work (units)
Car A	40	100	50
Car B	80	150	80
Car C	100	250	100

Labour costs £2 per hour, units of material cost 10s. each and units of subcontracted work cost £1 per unit. Find the total cost of manufacturing 3000, 2000 and 1000 vehicles of type A, B and C respectively.

14. Find the latent roots and proper vectors of the matrix

$$\begin{pmatrix} 4 & 1 \\ 2 & 3 \end{pmatrix}$$

EXERCISES ON VECTORS

1. Show that the set of all vectors is an infinite commutative group under addition. What is the identity element?

2. $ABCD$ is a quadrilateral. M is the midpoint of BD and N the midpoint of AC. Show that

$$\overrightarrow{AB} + \overrightarrow{AD} + \overrightarrow{CB} + \overrightarrow{CD} = 4\overrightarrow{NM}$$

3. $ABCD$ is a quadrilateral. By taking the position vectors of A, B, C, D relative to a fixed point O as **a**, **b**, **c**, **d**, obtain the position vectors of the midpoints of the sides and hence show that these midpoints form the vertices of a parallelogram.

4. If the diagonals of $ABCD$ in Question 3 are equal, prove that the midpoints of the sides form the vertices of a rhombus.

5. ABC is right angled at B and X is the midpoint of AC. Taking the position vectors of A and C relative to B as **a** and **c**, express $\overrightarrow{BX}$ and $\overrightarrow{AX}$ in terms of **a** and **c** and hence show that $AX = BX$.

6. In the parallelogram $ABCD$, P and Q lie on the diagonal BD so that $BQ = PD$. Show, by a vector method, that $AQCP$ is a parallelogram.

7. Circles described on AB, AC as diameters meet again at P. Prove, by a vector method, that B, P, C are collinear.

(*Hint:* Show that the vector product of $\overrightarrow{BP}$ and $\overrightarrow{BC}$ is zero.)

8. Two circles touch externally at A. A common tangent touches the circles at B and C respectively. Prove, by a vector method, that $\angle BAC$ is a right angle.

9. Two circles touch externally at A. LR and MS are parallel diameters of the two circles. Prove, by a vector method, that LAS is a straight line.

10. In triangles ABC, XYZ, AX, BY and CZ, produced if necessary, meet at L. AB, XY intersect at P; BC, YZ meet at Q and AC, XZ meet at R. Prove that P, Q and R are collinear.

(*Hint:* This is known as Desargues' theorem. Let A, B, C, X, Y, Z, etc., have position vectors **a**, **b**, **c**, **x**, **y**, **z**. Then

$$\lambda\mathbf{a}+\lambda'\mathbf{x} = \mu\mathbf{b}+\mu'\mathbf{y} = \eta\mathbf{c}+\eta'\mathbf{z} = \mathbf{l}$$

wher
$$\lambda+\lambda' = \mu+\mu' = \eta+\eta' = 1$$

$$\therefore \frac{\lambda\mathbf{a}-\mu\mathbf{b}}{\lambda-\mu} = \frac{\lambda'\mathbf{x}-\mu'\mathbf{y}}{\lambda'-\mu'} = \mathbf{p}$$

By obtaining similar expressions for **q** and **r**, show that

$$l\mathbf{p}+m\mathbf{q}+n\mathbf{r} = 0$$
$$l+m+n = 0$$

which are the conditions for the collinearity of P, Q and R.)

11. A and B are the points $(1, 2)$ $(5, 5)$ respectively and O is the origin of the rectangular coordinates. Writing $\overrightarrow{OA} = \mathbf{i}+2\mathbf{j}$, $\overrightarrow{OB} = 5\mathbf{i}+5\mathbf{j}$, obtain, by vector methods,

(i) the length of AB,
(ii) the point dividing AB in the ratio $3:1$,
(iii) the angle AOB,
(iv) the length of the perpendicular from O on to BA produced.

12. A, B, C, D are the points $(4, 6, 3)$, $(3, 1, 1)$, $(5, 2, 0)$ and $(2, 1, 4)$ respectively. Writing $\overrightarrow{OA} = 4\mathbf{i}+6\mathbf{j}+3\mathbf{k}$, etc., show, by a vector method, that AB and CD are perpendicular. If E is the point $(5, 11, 5)$ show that A, B, E are collinear and that A is the midpoint of BE.

13. A, B, C are the points $(0, 2, 5)$, $(2, 0, 7)$, $(1, 2, 0)$ and O is the origin. Find the resultant of the forces $\overrightarrow{OA}$, $\overrightarrow{OB}$, $\overrightarrow{OC}$.

14. $A = \mathbf{i}+\mathbf{j}+\mathbf{k}$, $B = \mathbf{i}-\mathbf{j}+\mathbf{k}$, $C = \mathbf{i}+\mathbf{j}-\mathbf{k}$.
Evaluate $\mathbf{A}.(\mathbf{B}\times\mathbf{C})$ and $\mathbf{A}\times(\mathbf{B}\times\mathbf{C})$.

15. If the three concurrent edges of a parallelepiped are represented by vectors **a**, **b**, **c**, prove that the scalar triple product $\mathbf{a}.(\mathbf{b}\times\mathbf{c})$ is a measure of its volume.

16. If **a**, **b**, **c** are the sides of a triangle taken in order and considered as vector quantities, then $\mathbf{a}+\mathbf{b}+\mathbf{c} = 0$. By multiplying both sides of this equation vectorially by **a** prove the sine rule.

17. Fig. 130 shows a one foot cube. The point of application of the force $\overrightarrow{AB'}$ (lb weight) is displaced from A to C'. Find the work done and the moment of the force $\overrightarrow{AB'}$ about the point C.

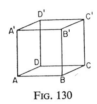

Fig. 130

EXERCISES ON LINEAR PROGRAMMING

1. A rectangular board has to be constructed so that its perimeter is less than 20 ft and greater than 12 ft. The ratio of the adjacent sides must be greater than 1:1 and less than 2:1. What integral dimensions satisfy these requirements?

2. A firm which supplies raw material in bulk has two depots D_1 and D_2 in a certain area. D_1 and D_2 currently stock 140 and 40 tons of material respectively. Two customers C_1 and C_2 place orders for 100 and 50 tons respectively. The cost of transport is a fixed amount per ton per mile. C_1 is 60 miles from D_1 and 30 miles from D_2; C_2 is 80 miles from D_1 and 20 miles from D_2. How should the deliveries be made in order to minimize transport costs?

(*Hint:* Let C_1, C_2 receive x, y tons respectively from D_1.)

3. 10 g of alloy A contain 2 g of copper, 1 g of zinc and 1 g of lead; 10 g of alloy B contain 1 g of copper, 1 g of zinc and 3 g of lead. It is required to produce a mixture of these alloys which contains at least 10 g of copper, 8 g of zinc and 12 g of lead. Alloy B costs $1\frac{1}{2}$ times as much per kilogramme as alloy A. Find the amounts of alloy A and alloy B which must be mixed in order to satisfy these conditions in the cheapest possible way.

(*Hint:* Consider a mixture of x, y dekagrams of A and B respectively.)

4. A small manufacturer employs 5 skilled men and 10 semi-skilled men and makes an article in two qualities, a de luxe model and an ordinary model. The making of a de luxe model requires 2 hr work by a skilled man and 2 hr work by a semi-skilled man; the ordinary model requires 1 hr work by a skilled man and 3 hr work by a semi-skilled man. By union rules no man may work more than 8 hr per day. The manufacturer's clear profit on the de luxe model is 10s. and on the ordinary model 8s. How many of each type should he make in order to maximize his total daily profit?

(*Hint:* Assume that he makes x ordinary and y de luxe models each day.)

EXERCISES ON STATISTICS

1. Represent by means of a pie-chart the following data. The area of land (including inland water) in the United Kingdom is, in units of 1000 acres, as follows:

England 32,212, Wales 5,130, Scotland 19,463.

2. Represent by means of a bar-chart the following data. The daily average number of hours of sunshine each month in England and Wales is as follows:

Jan. 1·51	Feb. 2·31	March 3·76	April 5·02
May 6·09	June 6·70	July 5·82	Aug. 5·47
Sept. 4·40	Oct. 3·18	Nov. 1·89	Dec. 1·34

3. The weights of 100 students are as follows:

4–6 stones	1	12–14 stones	19
6–8 stones	10	14–16 stones	3
8–10 stones	28	16–18 stones	1
10–12 stones	37	18–20 stones	1

Draw a histogram. State the modal and median values. Calculate the mean, mean deviation and standard deviation.

4. Thirty students carried out experiments to determine Joule's mechanical equivalent of heat. The results, corrected to two significant figures, were as follows:

Value obtained	3·8	3·9	4·0	4·1	4·2	4·3	4·4	4·5	4·6
Frequency	1	1	6	6	7	5	2	1	1

Determine the mean value, the mean deviation and the standard deviation from the mean.

5. Find the standard deviation of the set of integers

$$\{0, 1, 2, \ldots, 100\}$$

6. For the set $\{2, 4, 6, 8, 10\}$ show that the mean of the means of all possible pairs of elements is equal to the mean of the set itself. Does the same apply to the set $\{2, 4, 8, 16, 32, 64\}$?

7. The possible results of tossing three coins are as follows:

HHH, HHT, HTH, THH, HTT, THT, TTH, TTT.

Thus we have:

No. of heads	0	1	2	3
Frequency	1	3	3	1

Investigate the result of tossing 4, 5 coins. Can you generalize the result?

8. Find the "best straight line" of the form $y = mx$ which fits the following set of readings:

x	1	2	3	4	5	6
y	0·5	1·2	1·3	2·1	2·4	3

9. The annual export figures for two types of vehicle A and B to a certain European country were as follows:

	1958	1959	1960	1961	1962	1963
A	250	280	340	310	320	300
B	160	170	210	220	230	210

Calculate the coefficient of correlation.

10. The marks obtained by 5 pupils in English tests were as follows:

Eng. Language	9	4	2	7	8
Eng. Literature	6	5	4	8	7

Calculate the coefficient of correlation.

If, instead of marks, we record positions or ranks, we have:

Eng. Language	1	4	5	3	2
Eng. Literature	3	4	5	1	2

and differences in ranks are

d	2	0	0	2	0

Evaluate $r = 1 - \dfrac{6 \sum d^2}{N(N^2 - 1)}$, where N is the number of pairs of observations.

(r is known as the coefficient of rank correlation and does not usually vary much from the normal coefficient of correlation. The reader may care to evaluate both coefficients in Question 9.)

HINTS AND SOLUTIONS TO THE EXERCISES

SETS

1. (a) $\mathscr{E}$. (b) {Tuesday, Friday}. (c) {Wednesday}.
(d) {Monday, Thursday}. (e) P. (f) $\mathscr{E}$.
(g) $\emptyset$. (h) {Monday, Wednesday, Thursday}.
(i) $\emptyset$. (j) {Monday, Wednesday, Thursday}.
(g), (i) and (h), (j) are equal.

2. (a) $\{D\}$. (b) $\{B, C, D, I\}$. (c) $\{A, D, E, H\}$.
(d) $\{D\}$. (e) $\mathscr{E}$. (f) $\{A, D, E, H\}$. (g) $\{A, D, E, H\}$.
(h) $\{A, B, C, D, E, H, I\}$. (i) $\{A, B, C, D, E, H, I\}$.

(a) and (d); (f) and (g); (h) and (i). The last two pairs are always equal.

3. Both equal to $\{c\}$. Second two expressions unequal.

4. (a) $\{2\}$. (b) $\emptyset$. (c) $\mathscr{E}$. (d) $\emptyset$.
(e) $\{2, 3, 4, \ldots\}$. (f) $\mathscr{E}$. (g) $\emptyset$.

5. (a) $\mathscr{E}$. (b) $\{2\}$. (c) $\{2\}$. (d) $\{3, 4\}$. (e) $\{3\}$. (f) $\emptyset$.
(g) $\mathscr{E}$.

6. (a) $\{(1, 1)\ (2, 2)\ (3, 3)\ (4, 4)\}$.
  (b) $\{(1, 2)\ (2, 3)\ (3, 4)\}$.
  (c) $\{(1, 1)\ (1, 2)\ (1, 3)\ (1, 4)\ (2, 2)\ (2, 3)\ (2, 4)\ (3, 3)\ (3, 4)\}$.
  (d) $\{(2, 1)\ (3, 1)\ (3, 2)\ (4, 1)\ (4, 2)\ (4, 3)\}$.
  (e) $\{(4, 2)\}$.
  (f) $\{(2, 1)\ (2, 2)\ (2, 3)\ (2, 4)\ (1, 4)\ (2, 4)\ (3, 4)\ (4, 4)\}$.
  (g) $\{(2, 2)\}$.
  (h) $\{(1, 1)\ (1, 2)\ (1, 3)\ (1, 4)\ (2, 1)\ (2, 2)\ (2, 3)\ (2, 4)$
       $(3, 1)\ (3, 2)\ (3, 3)\ (4, 1)\ (4, 2)\}$.

7.

Fig. 131

Fig. 132

Fig. 133

Fig. 134

Fig. 135

Fig. 136

Fig. 137

8. i.e. 2 take part in all three activities.

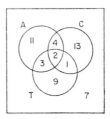

Fig. 138

9. 18 cars had faulty lights. 28 cars had only one fault.

Fig. 139

10. (a) f_4. (b) f_3 (c) f_1. (d) f_4. (e) f_1. (f) f_1.

LOGIC AND BOOLEAN ALGEBRA

1. (a) *F.* (b) *F.* (c) *T.* (d) *F.* (e) *F.* (f) *F.* (g) *F.* (h) *F.* (i) *F.* (j) *T.* (k) *F.* (l) Not valid but true. Squares.

2. (a) White cows do not wear bells. (It is untrue to say that Farmer Jones's cows do not wear bells.)
   (b) Some red-haired people are not bad-tempered.
   (c) Some old things are beautiful.

3. Boys must not wear a tie and must wear a shirt.

4. (i) I like tennis and I am a good cricketer.
   (ii) I am either a good cricketer or a good tennis player (or both).
   (iii) I like tennis and I am not a good cricketer.

HINTS AND SOLUTIONS TO EXERCISES

(iv) Either I do not like tennis or I am a good cricketer (or both).
(v) I do not like tennis and I am not a good cricketer.
(vi) Since I like tennis I am a good cricketer.

5.

(i)

p	$\sim p$	$p \vee \sim p$
T	F	T
F	T	T

(ii)

p	q	$\sim q$	$p \vee \sim q$
T	T	F	T
T	F	T	T
F	T	F	F
F	F	T	T

(iii)

p	q	$p \wedge \sim q$
T	T	F
T	F	T
F	T	F
F	F	F

(iv)

p	q	$\sim p$	$\sim p \wedge q$
T	T	F	F
T	F	F	F
F	T	T	T
F	F	T	F

(v)

p	q	$p \wedge \sim q$	$\sim p \wedge q$	$(p \wedge \sim q) \vee (\sim p \wedge q)$
T	T	F	F	F
T	F	T	F	T
F	T	F	T	T
F	F	F	F	F

6. (a) $(x'+y')(x+y')(x+y)$ simplifies to xy'.
   (b) $(x+yz)+(x+yz)(x+yz)$ simplifies to $x+yz$.
   (c) $(x+y)(x+z)(x+y)(x+z)$ simplifies to $x+yz$.
   (d) $(a+b+c+xy)(a+b+c+x'+y')$ simplies to $a+b+c$.
   (e) $(a+b)(a+c)(a+b'+c')$ simplifies to a.
   (f) $(a+bc)(a+b'+c')(a+b+c)(a+b+c')$ simplifies to $a+ab$ or a.
   (g) $abc+a'b'c+a'bc'+ab'c'$.

240 MODERN MATHEMATICS IN SECONDARY SCHOOLS

(h) $(p+q+r+pqr)(p+q+r+p'+q'+r')(p+q+r')(p+q')$ simplifies to p.

(i) $\{[(p+q)r+pq]r+pr\}s+prs$ simplifies to $rs(p+q)$.

7.

	a	b	c	$a'b'c$	$a'bc'$	$ab'c'$	abc	$abc+a'b'c+$ $+a'bc'+ab'c'$
All on	1	1	1	0	0	0	1	**1**
Any one change no light	1	1	0	0	0	0	0	0
	1	0	1	0	0	0	0	0
	0	1	1	0	0	0	0	0
Any successive change light	1	0	0	0	0	1	0	1
	0	1	0	0	1	0	0	1
	0	0	1	1	0	0	0	1
Further change no light	0	0	0	0	0	0	0	0

This is the closure table for a circuit in which three switches independently control one light.

8.

p	q	r	Required current
0	1	1	0
0	0	1	0
0	1	0	0
0	0	0	0
1	1	1	1
1	0	1	1
1	1	0	1
1	0	0	0

fuse intact $\equiv 1$

fuse blown $\equiv 0$

current flows $\equiv 1$

machine stops $\equiv 0$ (no current)

Current must flow for $p \wedge q \wedge r$ or $(\vee) p \wedge \sim q \wedge r$ or $p \wedge q \wedge \sim r$.

Therefore Boolean function is

$$(p \wedge q \wedge r) \vee (p \wedge \sim q \wedge r) \vee (p \wedge q \wedge \sim r)$$

or $\qquad pqr + pq'r + pqr'$

This is equal to $\qquad pr(q+q') + pqr'$

$$= pr + pqr' \quad (q+q' = 1)$$
$$= p(r + qr')$$
$$= p(r+r')(r+q)$$
$$= p(r+q) \quad (r+r' = 1)$$

Therefore required fuse circuit is as shown in Fig. 140.

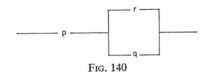

Fig. 140

9.

	a	b	c	d	e	Buzzer
Cases		1	1	1	1	1
where		1	1	1	0	1
buzzer		1	1	0	1	1
must		1	0	1	1	1
sound		0	1	1	1	1
	1	1	1	0	0	1
	1	1	0	1	0	1
	1	1	0	0	1	1
	1	0	1	1	0	1
	1	0	1	0	1	1
	1	0	0	1	1	1

Current flows in these cases:

$\qquad bcde$ or $\quad bcde'$ or $\quad bcd'e$ or $\quad bc'de$ or $\quad b'cde$

and

$abcd'e'$ or $abc'de'$ or $abc'd'e$ or $ab'cde'$ or $ab'cd'e$ or $ab'c'de$.

Therefore Boolean function is

$$bcde + \sum b'cde + a \sum b'c'de$$

Which after simplification gives the circuit shown in Fig. 141.

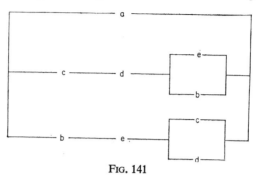

FIG. 141

GROUPS

1, 2. Denote $\{e^{0.i\pi/3}, e^{1.i\pi/3}, e^{2.i\pi/3}, \ldots, e^{5.i\pi/3}\}$ by $\{0, 1, 2, 3, \ldots, 5\}$. Then both tables are:

+	0	1	2	3	4	5	(mod 6)
0	0	1	2	3	4	5	
1	1	2	3	4	5	0	
2	2	3	4	5	0	1	
3	3	4	5	0	1	2	
4	4	5	0	1	2	3	
5	5	0	1	2	3	4	

They are commutative and isomorphic under the mapping

$$e^{n.i\pi/3} \longleftrightarrow n \quad (n = 0, 1, \ldots, 5)$$

3. Any cyclic group of order n is isomorphic with the group of integers modulo n under addition.

4. If $I = \begin{pmatrix} 1 & 2 \\ 1 & 2 \end{pmatrix}$ $P = \begin{pmatrix} 1 & 2 \\ 2 & 1 \end{pmatrix}$ we have

	I	P
I	I	P
P	P	I

and all groups of order 2 must be of this form.

HINTS AND SOLUTIONS TO EXERCISES

5.

	f_1	f_2	f_3	f_4	f_5	f_6	2nd function
f_1	f_1	f_2	f_3	f_4	f_5	f_6	
f_2	f_2	f_3	f_1	f_6	f_4	f_5	
f_3	f_3	f_1	f_2	f_5	f_6	f_4	
f_4	f_4	f_5	f_6	f_1	f_2	f_3	
f_5	f_5	f_6	f_4	f_3	f_1	f_2	
f_6	f_6	f_4	f_5	f_2	f_3	f_1	

1st function (row labels at left)

6. $\{0, 3\}$ and $\{0, 2, 4\}$.

7. $\{p_0, p_1, p_2\}$ as in Chapter 3.

8. $\{f_1, f_2\}, \{f_1, f_3\}, \{f_1, f_4\}$ are respectively isomorphic with $\{1, 3\}, \{1, 5\}, \{1, 7\}$ mod 8 under multiplication.

9. The matrices are each of the form $\begin{pmatrix} \cos\dfrac{-n\pi}{3} & -\sin\dfrac{n\pi}{3} \\ \sin\dfrac{n\pi}{3} & \cos\dfrac{n\pi}{3} \end{pmatrix}$

$(n = 0, 1, 2, \ldots, 5)$. They therefore create successive anticlockwise rotations of $0, \dfrac{\pi}{3}, \dfrac{2\pi}{3}, \ldots, \dfrac{5\pi}{3}$ in the Cartesian plane. This is precisely the operation performed by $e^0, e^{i\pi/3}, \ldots, e^{5i\pi/3}$, in the Argand plane (Question 1). Thus this group of matrices must be isomorphic with the groups in Questions 1 and 2.

MATRICES

1. P, Q respectively transform into:

(a) $(1, 1)$ $(2, 2)$ PQ unchanged—the "identity" transformation.

(b) $(1, -1)$ $(2, -2)$ $PQ \to$ its reflection in the X-axis.

(c) $(-1, 1)$ $(-2, 2)$ $PQ \to$ its reflection in the Y-axis.

(d) $(0, \sqrt{2})$ $(0, 2\sqrt{2})$ PQ is rotated about 0 through $\tfrac{1}{4}\pi$ radians.

(e) $(0, 2)$ $(0, 4)$ $\overrightarrow{OP}, \overrightarrow{OQ}$ and $\overrightarrow{PQ}$ increased in the ratio $\sqrt{2}:1$ and the result rotated about 0 through $\tfrac{1}{4}\pi$ radians.

(f) (2, 2) (4, 4) $P \to Q$, $Q \to (4, 4)$ i.e. PQ is translated along $\overrightarrow{PQ}$ until P coincides with Q, the segment is then enlarged in the ratio 2:1.

(g) (3, 3) (6, 6) PQ translated along $\overrightarrow{PQ}$ until $P \to (3, 3)$, the segment is then enlarged in the ratio 3:1.

2. $$\begin{pmatrix} x_3 \\ y_3 \end{pmatrix} = \begin{pmatrix} 1 & 1 \\ 1 & -1 \end{pmatrix} \cdot \begin{pmatrix} 1 & 2 \\ 2 & -1 \end{pmatrix} \cdot \begin{pmatrix} 2 & -1 \\ 1 & 2 \end{pmatrix} \cdot \begin{pmatrix} x_0 \\ y_0 \end{pmatrix}$$
$$= \begin{pmatrix} 7 & -1 \\ 1 & 7 \end{pmatrix} \cdot \begin{pmatrix} x_0 \\ y_0 \end{pmatrix}$$

or
$$x_3 = 7x_0 - y_0$$
$$y_3 = x_0 + 7y_0$$

3. (a) $\begin{pmatrix} 2 & 0 \\ 0 & 2 \end{pmatrix}$. (b) $\begin{pmatrix} 3 & 0 \\ 0 & 3 \end{pmatrix}$. (c) $\begin{pmatrix} 1 & -\sqrt{3} \\ \sqrt{3} & 1 \end{pmatrix}$.

(d) $\begin{pmatrix} 1 & 0 \\ 0 & -1 \end{pmatrix}$. (e) $\begin{pmatrix} -1 & 0 \\ 0 & 1 \end{pmatrix}$.

(f) $\begin{pmatrix} x_1 \\ y_1 \end{pmatrix} = \begin{pmatrix} 1 & 0 \\ 0 & 1 \end{pmatrix} \cdot \begin{pmatrix} x_0 \\ y_0 \end{pmatrix} + \begin{pmatrix} 3 \\ 4 \end{pmatrix}$.

Put $(x_0, y_0) = (0, 0)$ (1, 1).

4. $(0, 0) \to (0, 0)$ $\begin{matrix} (1, 0) \searrow \\ (0, 1) \nearrow \end{matrix} (1, 1)$ $(1, 1) \to (2, 2)$

All points on and inside the square are mapped on to the line
$$y = x, \quad 0 \leq x \leq 2.$$

5. $X^2 + \dfrac{Y^2}{K^2} = 1$. (a) $\pi ab = \frac{1}{2}\pi a^2$ (b) $k^2 = 1^2(1 - e^2)$

$\therefore k = \frac{1}{2}$ $\therefore k = \dfrac{\sqrt{3}}{2}$

6. (a) $\begin{pmatrix} 20 & 8 \\ 17 & 17 \end{pmatrix}$. (b) $\begin{pmatrix} 17 & 9 & 16 \\ 3 & 6 & 9 \\ 7 & 9 & 14 \end{pmatrix}$.

HINTS AND SOLUTIONS TO EXERCISES   245

7. (a) $\begin{pmatrix} 8 & 17 \\ 9 & 16 \end{pmatrix}.$ (b) $\begin{pmatrix} 6 & 19 \\ 7 & 18 \end{pmatrix}.$ (c) $\begin{pmatrix} 13 & 14 \\ 14 & 17 \end{pmatrix}.$ (d) $\begin{pmatrix} 17 & 14 \\ 14 & 13 \end{pmatrix}.$
(e) $\begin{pmatrix} 6 & 7 \\ 19 & 18 \end{pmatrix}.$ (f) $\begin{pmatrix} 6 & 7 \\ 19 & 18 \end{pmatrix}.$ (g) 25. (h) 25.

8. (a) $\begin{pmatrix} 3 & 0 & 0 \\ 0 & 2 & 0 \\ 0 & 0 & 6 \end{pmatrix}.$ (b) $\begin{pmatrix} 3 & 0 & 0 \\ 0 & 2 & 0 \\ 0 & 0 & 6 \end{pmatrix}.$ (c) $\begin{pmatrix} 3 & 0 & 0 \\ 0 & 2 & 0 \\ 0 & 0 & 6 \end{pmatrix}.$
(d) $\begin{pmatrix} 3 & 0 & 0 \\ 0 & 2 & 0 \\ 0 & 0 & 6 \end{pmatrix}.$ (e) 36. (f) 36.

9. (a) $\begin{pmatrix} 0 & 0 \\ 0 & 0 \end{pmatrix}.$ (b) $\begin{pmatrix} -1 & -1 \\ 1 & 1 \end{pmatrix}.$

10. $\begin{pmatrix} 2 & 6 & 10 \\ 6 & 10 & 14 \\ 10 & 14 & 18 \end{pmatrix} + \begin{pmatrix} 0 & -2 & -4 \\ 2 & 0 & -2 \\ 4 & 2 & 0 \end{pmatrix}.$

11. $\frac{1}{10}\begin{pmatrix} 1 & 2 \\ -3 & 4 \end{pmatrix}.$ (a) $x = \frac{1}{10}(p+2q), y = \frac{1}{10}(-3p+4q).$
(b) $x = 4, y = 2.$ (c) $x = 1, y = 2.$
(d) $x = 1, y = -\frac{1}{2}.$

12. $x = 2, \quad y = 1, \quad z = 1.$

13. £1,595,000.

14. $\lambda = 2$ or 5; for $\lambda = 2$ $(-k, 2k) \to (-2k, 4k)$
for $\lambda = 5$ $(h, h) \to (5h, 5h).$

VECTORS

1. Identity element is the null or zero vector. The inverse of **a** is $-\mathbf{a}$. We have proved that vector addition is commutative and associative and, of course, vector addition is a binary operation.

2. L.H.S. $= 2\overrightarrow{AM} + 2\overrightarrow{CM} = 4\overrightarrow{NM}.$

3. $\frac{a+b}{2}, \frac{b+c}{2}$, etc. Vector joining these midpoints is $\frac{c-a}{2}$. Vector joining $\frac{a+d}{2}, \frac{c+d}{2}$ is $\frac{c-a}{2}$, i.e. opposite sides parallel, etc.

5. $\overrightarrow{AX} = \frac{1}{2}(c-a)$. $\overrightarrow{BX} = \frac{1}{2}(c+a)$ and $|c-a| = |c+a|$ since $c \cdot a = 0$ ($\angle B = 90°$).

6. Put $\overrightarrow{DA} = \overrightarrow{CB} = y$, $\overrightarrow{DP} = \overrightarrow{QB} = z$, then $\overrightarrow{PA} = \overrightarrow{CQ} = y-z$
$\therefore |PA| = |CQ|$ and PA is parallel to CQ.

8. Put $\overrightarrow{AD} = T$, $\overrightarrow{BD} = \overrightarrow{DC} = t$ then
$$\overrightarrow{BA} \cdot \overrightarrow{AC} = (t-T) \cdot (T+t) = 0.$$

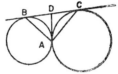

Fig. 142

9. Let unit vector r act along line of centres and unit vector b act along $\overrightarrow{LR}$. Then $\overrightarrow{LA} \times \overrightarrow{AS} = (nr + nb) \times (mr + mb) = 0$.

11. (i) 5. (ii) $(4, 4\frac{1}{4})$.

(iii) $\quad |\overrightarrow{BO} \times \overrightarrow{BA}| = BO \cdot BA \sin \angle ABO$

$\therefore \cos \angle AOB = \frac{(i+2j) \cdot (5i+5j)}{\sqrt{5} \times 5\sqrt{2}} = \frac{3\sqrt{10}}{10}$

(iv) $\overrightarrow{AB} = \overrightarrow{OB} - \overrightarrow{OA} = 4i + 3j$

$|\overrightarrow{BO} \times \overrightarrow{BA}| = BO \cdot BA \sin ABO = BA \cdot p = 5p$
$= 5|(i+j) \times (4i+3j)|$
$\therefore p = 1$.

13. 13 acting along the line with direction cosines
$(\frac{3}{13}, \frac{4}{13}, \frac{12}{13})$.

14. 4; $2(\mathbf{k}-\mathbf{j})$.

16. $\mathbf{a}+\mathbf{b}+\mathbf{c} = 0$
$\mathbf{a}\times\mathbf{a}+\mathbf{a}\times\mathbf{b}+\mathbf{a}\times\mathbf{c} = 0$
$\therefore ab \sin C.\mathbf{n}-ac \sin B.\mathbf{n} = 0$
$\therefore \dfrac{b}{\sin B} = \dfrac{c}{\sin C}$ etc.

17. W.D. $= (\mathbf{i}+\mathbf{k}).(\mathbf{i}+\mathbf{j}+\mathbf{k}) = 2$ ft-lb
moment required $= |(\mathbf{i}+\mathbf{k})\times(\mathbf{i}+\mathbf{j})| = |\mathbf{k}+\mathbf{j}-\mathbf{i}|$
$= \sqrt{3}$ lb-ft.

LINEAR PROGRAMMING

1. $3'\times 4'$; $3'\times 5'$; $4'\times 5'$.

2. $x = 100$, $y = 10$.

The inequalities are: $x+y \leqq 140$, $\quad 0 \leqq x \leqq 100$,
$\qquad\qquad\qquad\qquad x+y \geqq 110$, $\quad 0 \leqq y \leqq 50$.

3. $x = 6$, $y = 2$.

The inequality matrix is $\begin{pmatrix}2 & 1\\ 1 & 1\\ 1 & 3\end{pmatrix}.\begin{pmatrix}x\\ y\end{pmatrix} \geqq \begin{pmatrix}10\\ 8\\ 12\end{pmatrix}.$

The function to be minimized is $2x+3y$.

4. $x = 20$, $y = 10$.

Here $\begin{pmatrix}1 & 2\\ 3 & 2\end{pmatrix}.\begin{pmatrix}x\\ y\end{pmatrix} \leqq \begin{pmatrix}40\\ 80\end{pmatrix}$
$\qquad x > 0, \quad y > 0.$

The function to be maximized is $5y+4x$.

STATISTICS

3. The mode and median lie in the range 10–12 stones. The mean value is $\simeq 10.6$ stones (10·62).

The mean deviation is $\simeq 1.74$ stones (1·7436).
The standard deviation is $\simeq 2.28$ (2·276).

4. Mean value $= 4\cdot 2$ to 2 significant figures ($4\cdot 17$).
Mean deviation from $4\cdot 2 = 0\cdot 14$ to 2 significant figures ($0\cdot 137$).
Standard deviation from $4\cdot 2 = 0\cdot 18$ to 2 significant figures ($0\cdot 176$).

5. $29\cdot 15$.

6. Yes. Mean of first set $= 6$; mean of second $= 21$.

7. For 0, 1, 2, 3, 4 heads.
$f = 1, 4, 6, 4, 1$.
For 0, 1, 2, 3, 4, 5 heads
$f = 1, 5, 10, 10, 5, 1$.

The values of f for n heads are (in order) the binomial coefficients in the expansion of $(1+x)^n$.

8. $y = 0\cdot 5x$ (approximately).

9. $0\cdot 85$.   Coefficient of rank correlation $= 0\cdot 76$.

10. Coefficient of correlation $= 0\cdot 76$.
Coefficient of rank correlation $= 0\cdot 6$.

INDEX

Abelian group 55, 58
Angular velocity 125
Argand plane 55, 56, 68
Associative property 57

Bar-chart 159
Binary
  adder 214
  fractions 213
  operation 46
  scale 213
  slide rule 212
Bisector theorem 111
Boolean
  algebra 33
  function 37, 39, 41-4

Centroid 110
Characteristic equation 95
Chinese multiplication 188
Closed half-plane 135
Closure 57
Closure tables 30, 36, 38, 40
Collinear points 103
Commutative 58
Complement 3
Complex
  numbers 55, 68
  roots 49
Compound statements 28–30
Conjunction 26, 32
Convex sets 136
Coplanar vectors 109
Correlation 173
  coefficient of 176
  coefficient of rank 235
Cosine rule 116
Cross-ration 226

Desargues' theorem 231
Disjoint sets 3
Disjunction 26, 32
Domain 12
Dual property 16

Eigenvector 96
Element 2
Empty set 3
Enlargement 74, 129
Equal sets 6
Equivalence 26, 101
Exclusive disjunction 29
Extreme points 136

Farey series 202
Fibonacci sequence 204
Fictitious mean 166
Fractions 101, 201
  binary 213
Frequency
  curve 164
  polygon 161
Function 10, 32

Golden
  number 207
  section 206
  spiral 207
Group
  of primes 63
  of the rectangle 53
  of the square 54
  of the triangle 61, 64
Groups 46
  cyclic 62
  definition of 57

Groups
  isomorphic 62, 63
  permutation 64
  substitution 62

Half-plane 135
Half-turns 50, 51
Histogram 162
Hypotheses 22

Identity
  element 58
  operation 51
Implication 26
Indices 210
Inequality 7, 135
Intersection 3, 135–6, 152
Inverse 14, 51, 58
Isomorphic 62
Iterative process 84

Latent roots 97
Latin squares 193
Lattice 8, 132
Laws of sets 15
Least squares 172
Linear
  programming 136
  transformation 75
Listing 2
Logarithms 210
Logic 33
Logically true 29

Magic squares 189
Mapping 11, 98, 226
Matrices 70
  addition of 81
  algebra of 81, 82
  antisymmetric 83
  diagonal 80
  equal 78
  inverse of 86, 88, 91, 129
  magic 192
  multiplication 72, 77, 81, 116
  null 80
  singular 70
  square 70
  symmetric 83
  transposed 78, 80
  unit 73
Mean 165
  deviation 169
Median (value) 167
Medians (of triangle) 49, 109
Menelaus' theorem 127
Midpoint theorem 107
Mode 167
Modulo (arithmetic) 58
Moment of a force 125
Morse code 214
Motion geometry 129

Negation 26, 32
Nim (game of) 215
Normal distribution curve 165, 180

One-to-one correspondence 11, 14, 63
Open half-plane 135
Ordered pair 8

Parallelogram law 99
Pascal's triangle 195
Permutation 64, 65
Pie-chart 159
Powers of numbers 200
Premises 22
Prime numbers 198, 199
Probability 182
Product
  of mappings 13
  set 10
Proper subset 3
Pythagoras' theorem 116

Quarter-turns 69

Radius of gyration 170

Range 12, 168
Reductio ad absurdum 23
Reflection 129
Relation 10
Rotations 50–1, 68, 73, 129

Scalar product 47, 114
Semi-interquartile range 168
Set 1
Significance 179
Simultaneous equations 86
Sine rule 232
Slide rule 212
Solution sets 6
Square numbers 197
Standard deviation 169, 171
Statements 22
Statistics 156
Subgroups 55, 56, 57
Subset 2
Swing radius 170
Switch circuits 27, 36, 37, 38
Symmetry 47

Transformation 11, 70, 75, 76, 96
Translation 129
Transport problem 148
Triangular numbers 195
Truth
  set 26

  tables 28–31
  value 22

Union 3, 152
Universal set 3

Valid arguments 22
Variance 169
Vector
  area 124
  column 70
  equation of line 113
  free 99, 100
  modulus of 102
  null 102, 123
  position 105
  proper 96
  quantity 99
  row 70
  sliding 100
  tied 100
  unit 102, 111–2, 116
Vectors
  coplanar 109
  scalar product of 114
  vector product of 47, 123
Venn diagrams 4

Work done by force 124